# Project Management Professional Exam

## Review Guide

### Second Edition

# PMP®

# Project Management Professional Exam

## Review Guide

### Second Edition

Kim Heldman

Vanina Mangano

WITHDRAWN

John Wiley & Sons, Inc.

Senior Acquisitions Editor: Jeff Kellum
Development Editor: Sara Barry
Technical Editors: Terri Wagner and Brett Feddersen
Production Editor: Eric Charbonneau
Copy Editor: Judy Flynn
Editorial Manager: Pete Gaughan
Production Manager: Tim Tate
Vice President and Executive Group Publisher: Richard Swadley
Vice President and Publisher: Neil Edde
Media Project Manager 1: Laura Moss-Hollister
Media Associate Producer: Shawn Patrick
Media Quality Assurance: Josh Frank
Book Designers: Judy Fung and Bill Gibson
Compositor: Craig W. Johnson, Happenstance Type-O-Rama
Proofreaders: Brett Feddersen; Paul Sagan and Jeff Holt, Word One New York
Indexer: Ted Laux
Project Coordinator, Cover: Katherine Crocker
Cover Designer: Ryan Sneed

Copyright © 2011 by John Wiley & Sons, Inc., Indianapolis, Indiana

Published simultaneously in Canada

ISBN: 978-1-118-09391-7 (pbk)
ISBN: 978-1-118-16481-5 (ebk)
ISBN: 978-1-118-16482-2 (ebk)
ISBN: 978-1-118-16480-8 (ebk)

No part of this publication may be reproduced, stored in a retrieval system or transmitted in any form or by any means, electronic, mechanical, photocopying, recording, scanning or otherwise, except as permitted under Sections 107 or 108 of the 1976 United States Copyright Act, without either the prior written permission of the Publisher, or authorization through payment of the appropriate per-copy fee to the Copyright Clearance Center, 222 Rosewood Drive, Danvers, MA 01923, (978) 750-8400, fax (978) 646-8600. Requests to the Publisher for permission should be addressed to the Permissions Department, John Wiley & Sons, Inc., 111 River Street, Hoboken, NJ 07030, (201) 748-6011, fax (201) 748-6008, or online at `http://www.wiley.com/go/permissions`.

Limit of Liability/Disclaimer of Warranty: The publisher and the author make no representations or warranties with respect to the accuracy or completeness of the contents of this work and specifically disclaim all warranties, including without limitation warranties of fitness for a particular purpose. No warranty may be created or extended by sales or promotional materials. The advice and strategies contained herein may not be suitable for every situation. This work is sold with the understanding that the publisher is not engaged in rendering legal, accounting, or other professional services. If professional assistance is required, the services of a competent professional person should be sought. Neither the publisher nor the author shall be liable for damages arising herefrom. The fact that an organization or Web site is referred to in this work as a citation and/or a potential source of further information does not mean that the author or the publisher endorses the information the organization or Web site may provide or recommendations it may make. Further, readers should be aware that Internet Web sites listed in this work may have changed or disappeared between when this work was written and when it is read.

For general information on our other products and services or to obtain technical support, please contact our Customer Care Department within the U.S. at (877) 762-2974, outside the U.S. at (317) 572-3993 or fax (317) 572-4002.

Wiley also publishes its books in a variety of electronic formats and by print-on-demand. Not all content that is available in standard print versions of this book may appear or be packaged in all book formats. If you have purchased a version of this book that did not include media that is referenced by or accompanies a standard print version, you may request this media by visiting `http://booksupport.wiley.com`. For more information about Wiley products, visit us at `www.wiley.com`.

Library of Congress Cataloging-in-Publication Data is available from the publisher.

TRADEMARKS: Wiley, the Wiley logo, and the Sybex logo are trademarks or registered trademarks of John Wiley & Sons, Inc. and/or its affiliates, in the United States and other countries, and may not be used without written permission. PMP is a registered trademark of Project Management Institute, Inc. All other trademarks are the property of their respective owners. John Wiley & Sons, Inc., is not associated with any product or vendor mentioned in this book.

10 9 8 7 6 5 4 3 2 1

Dear Reader,

Thank you for choosing *PMP: Project Management Professional Exam Review Guide, Second Edition*. This book is part of a family of premium-quality Sybex books, all of which are written by outstanding authors who combine practical experience with a gift for teaching.

Sybex was founded in 1976. More than 30 years later, we're still committed to producing consistently exceptional books. With each of our titles, we're working hard to set a new standard for the industry. From the paper we print on, to the authors we work with, our goal is to bring you the best books available.

I hope you see all that reflected in these pages. I'd be very interested to hear your comments and get your feedback on how we're doing. Feel free to let me know what you think about this or any other Sybex book by sending me an email at `nedde@wiley.com`. If you think you've found a technical error in this book, please visit `http://sybex.custhelp.com`. Customer feedback is critical to our efforts at Sybex.

Best regards,

Neil Edde
Vice President and Publisher
Sybex, an Imprint of Wiley

*To BB, my forever love.*
*—Kim Heldman*

*To Al Smith, Jr., whose support and encouragement have given me the ability to do greater things.*
*—Vanina Mangano*

# Acknowledgments

I'd like to thank Vanina for an outstanding job writing this book. She was a lifesaver and bore the brunt of the work. I enjoyed seeing how she gave the content a fresh face, and her illustrations are terrific. (Don't tell her, but I may borrow a few of them for future reference.) It was a pleasure to work with Vanina. She was a real trouper through some of the rough starts and stops we had deciding what the content should look like. I hope we have the opportunity to work together again in the future.

I also echo Vanina's thanks to Jeff Kellum, senior acquisitions editor. This book, and the boxed set, was his brainchild. I always enjoy the opportunity to work with Jeff and all the great staff at Sybex.

I also want to thank Neil Edde, vice president and publisher at Sybex, for taking that leap of faith on a crazy project management study guide idea way back when. Thanks, Neil.

Sara Barry, development editor, is the best. She had some great suggestions that improved the content and kept us on track with deadlines just like a seasoned project manager.

Thanks to Terri Wagner, technical editor, for her help on this project. Terri is the president of Mentor Source, Inc., and conducts training classes all over the globe. Her perspective on the content and what she hears in real-life classroom situations helped us to make the content relevant and clear.

A special thank you to Brett Feddersen for the final technical edit of the book and his eagle-eye review. I have had the pleasure of working with Brett for many years, and he is an outstanding leader and project manager and a true friend.

A very big thanks goes to all of the instructors who use my books in their classrooms. I appreciate you choosing Sybex and my books to help your students master PMP concepts. Thank you also to all of the readers who choose this book to help them study for the PMP® exam.

—Kim Heldman

To start, thank you to the team at Sybex who devoted a great deal of effort toward making this review guide come together successfully. It is incredible to see the amount of teamwork and effort that goes into the making of a book. The process is certainly thorough, and there are many people not named here who were key to producing a solid product. Once again, thanks to all of you!

I'd like to thank Kim Heldman for the opportunity of working together on this book. It was wonderful to see that your live personality is just as dynamic, warm, and welcoming as your written voice. I had a blast getting to know your work in such a detailed way, and I am a bigger fan than ever!

Thank you to Jeff Kellum, our senior acquisitions editor, who saw the value of this book and whose great and supportive personality always comes through on the phone and in email. I appreciate you welcoming me to the Sybex family, and it is always a sincere pleasure working with you.

A tremendous thank you to Sara Barry, development editor. With your guidance and feedback, we were able to take this book to the next level, and I am thrilled with the result! You ensured that everything stayed on course, and your recommendations were valuable and instrumental in creating this finished product. It was absolutely wonderful having an opportunity to work with you.

Thank you to Terri Wagner, technical editor, whose sharp eyes made sure that we were on the ball. You were a key contributor in making sure the information was accurate and clearly communicated, and your knowledge of the field is very clear.

A special thanks to the individuals who are such a big part of my life and who have always impacted me in a magnificent way. This includes my family: Nicolas Mangano, Marysil Mangano, Nicolas Mangano, Jr., Carina Moncrief and her husband, Jonathan Moncrief, and my beautiful nieces—you mean everything to me! Thank you to Al Smith, Jr., my partner in all things, who is supportive and an inspiration to me—I enjoy all of our adventures and accomplishments together! And finally, thank you to Roshoud Brown, who always encouraged me to write and inspired me with his own words.

—Vanina Mangano

We both would like to thank Judy Flynn, copyeditor, who made sure grammar and spelling were picture perfect; Paul Sagan and Jeff Holt, proofreaders, for catching those last little "oops," and Eric Charbonneau, production editor, who made sure everything flowed through the production process. Thanks also to our compositor, Craig W. Johnson, and the indexer, Ted Laux. The book couldn't happen without them.

—The Authors

# About the Authors

**Kim Heldman,** MBA, PMP, is a senior IT director for the state of Colorado. Kim is responsible for overseeing technology services for four departments: Public Health and Environment, Natural Resources, Agriculture, and Local Affairs. She manages and oversees projects with IT components ranging from projects that are small in scope and budget to multimillion dollar, multiyear projects. She has over 20 years of experience in information technology project management. Kim has served in a senior leadership role for over 12 years and is regarded as a strategic visionary with an innate ability to collaborate with diverse groups and organizations, instill hope, improve morale, and lead her teams in achieving goals they never thought possible.

In addition to her project management experience, Kim has experience managing application development, web development, network operations, infrastructure, security, and customer service teams.

Kim is the author of the *PMP: Project Management Professional Exam Study Guide, 6th Edition* published by Sybex. Thousands of people worldwide have used the Study Guide in preparation for the PMP exam. Kim is also the author of *Project Management JumpStart, 3rd Edition* and *Project Manager's Spotlight on Risk Management* and coauthor of *CompTIA Project+* and *Excel 2007 for Project Managers*, all from Sybex. Kim has also published several articles and is currently working on a leadership book.

Kim continues to write on project management best practices and leadership topics, and she speaks frequently at conferences and events. You can contact Kim at `Kim.Heldman@comcast.net`. She personally answers all her email.

**Vanina Mangano** is an executive member of Never Limited LLC, which owns a portfolio of companies, including The PM Instructors. The PM Instructors specializes in PMI certification exam preparation material and training. Through The PM Instructors, Vanina has authored courseware material within the subject of project management, including the most recent release: *PMI Risk Management Professional Exam Preparation.*

Over the past decade, Vanina has specialized in working with and founding startup companies, ranging from overseas technology-driven companies to recreation and fitness companies. Her experience largely focuses on the areas of project management, operations, and business analysis. Prior to her current role, Vanina served as the VP of business development for BHNET Software Solutions, focusing in the onshore/offshore outsourcing model within the Asian and European countries.

As part of her contribution to the community, Vanina has participated in various committees and organizations. Currently, Vanina is serving as a core committee member of the 5th edition update project for *A Guide to the Project Management Body of Knowledge* (*PMBOK® Guide*) through PMI. Her role as integrated content and change control lead involves heading integration, editing, and data flow efforts. Vanina also served on the board of the Orange County network of Women in Technology International (WITI), participated in formal mentorship programs by acting as a mentor to tenured professional women in the technology industry, and has spoken at alumni business conferences.

Vanina holds a dual bachelor's degree from the University of California, Riverside, and holds the following credentials: Project Management Professional (PMP), PMI Risk Management Professional (PMI-RMP), PMI Scheduling Professional (PMI-SP), CompTIA Project+, and ITIL Foundation v3.

You may reach Vanina via email at `Vanina.Mangano@ThePM-Instructors.com`.

# Contents at a Glance

# Contents

# Introduction

Congratulations on your decision to pursue the Project Management Professional (PMP) credential, one of the most widely recognized credentials within the project management industry! The PMP credential is offered by the Project Management Institute (PMI), a not-for-profit organization with thousands of members across the globe. PMI has been a long-standing advocate and contributor to the project management industry and offers several credentials for those specializing in the field of project management.

This book is meant for anyone preparing to take the PMP certification exam as well as individuals who are looking to gain a better understanding of *A Guide to the Project Management Body of Knowledge, 4th Edition* (*PMBOK® Guide*). If you are studying for the Certified Associate in Project Management (CAPM®) exam, you may also find this book useful because the CAPM exam tests your knowledge of the *PMBOK® Guide* contents.

This review guide has been formatted to work hand in hand with *PMP: Project Management Professional Exam Study Guide, 6th Edition,* from Sybex. The Study Guide provides a more comprehensive review of the concepts included on the exam along with real-world examples. This review guide will reinforce these concepts and provide you with further explanation and a handy reference guide to the project management processes within the *PMBOK® Guide*. You'll find references to the Study Guide throughout this book, guiding you to where you may find additional information as needed. With all of these great resources at your fingertips, learning and understanding the *PMBOK® Guide*, along with other project management concepts, has certainly become easier!

## Book Structure

This book has been structured in a way that carefully follows the concepts of the *PMBOK® Guide*, allowing you to understand how a project is managed from beginning to end. For this reason, we will review the processes in the order of the process groups:

- Initiating
- Planning
- Executing
- Monitoring and Controlling
- Closing

We start by covering the project management framework and the *PMI Code of Ethics and Professional Conduct* and then move to a comprehensive review of the process groups. You'll find that each chapter offers a concise overview of each project management process and concept as well as the process inputs, tools and techniques, and outputs. This structure allows you to go back and reference terms, definitions, and descriptions at a glance.

## Overview of PMI Credentials

PMI offers several credentials within the field of project management, so whether you are an experienced professional or looking to enter into the project management field for the first time, you'll find something to meet your needs. You may hold one or multiple credentials concurrently.

Over the years, PMI has contributed to the project management body of knowledge by developing global standards used by thousands of project management professionals and organizations. In total, there are 11 standards grouped within the following categories:

- Projects
- Programs
- Profession
- Organizations
- People

Several credentials offered by PMI are based on the *PMBOK® Guide*, which is part of the Projects category. As of the publication date of this book, PMI offers five credentials. Let's briefly go through them:

**Project Management Professional (PMP)** You are most likely familiar with the PMP credential—after all, you purchased this book! But did you know that the PMP certification is the most widely and globally recognized project management certification? The PMP, along with several other credentials, validates your experience and knowledge of project management. This makes obtaining a PMP in itself a great achievement. The following requirements are necessary to apply for the PMP exam:

**Work Experience** The following work experience must have been accrued over the past eight consecutive years:

- If you have a bachelor's degree or the global equivalent: three years (36 months) of non-overlapping project management experience, totaling at least 4,500 hours
- If you have a high school diploma, associate's degree, or global equivalent: five years (60 months) of non-overlapping project management experience, totaling at least 7,500 hours

**Contact Hours** "Contact hours" refers to the number of qualified formal educational hours obtained that relate to project management. A total of 35 contact hours is required and must be completed before you submit your application.

**Certified Associate in Project Management (CAPM)** The CAPM credential is ideal for someone looking to enter the project management industry. You may meet the requirements through work experience *or* through formal project management education. If you do not currently have project management experience, you may apply if you have accumulated the requisite number of formal contact hours:

**Work Experience** 1,500 hours of formal project management experience

**Contact Hours** 23 contact hours of formal project management education

**Program Management Professional (PgMP®)** The PgMP credential is ideal for those who specialize in the area of program management or would like to highlight their experience of program management. A PMP is not required to obtain this or any other credential. You must meet the following requirements to apply for the PgMP exam:

**Work Experience** The following work experience must have been accrued over the past 15 consecutive years:

- If you have a bachelor's degree or global equivalent: four years of non-overlapping project management experience, totaling at least 6,000 hours, *and* four years of non-overlapping program management experience, totaling 6,000 hours
- If you have a high school diploma, associate's degree, or global equivalent: four years of non-overlapping project management experience, totaling at least 6,000 hours, *and* seven years of non-overlapping program management experience, totaling 10,500 hours

**PMI Risk Management Professional (PMI-RMP®)** The PMI-RMP credential is ideal for those who specialize in the area of risk management or would like to highlight their risk management experience. The following are the requirements to apply for the PMI-RMP exam:

**Work Experience** The following work experience must have been accrued over the past five consecutive years:

- If you have a bachelor's degree or global equivalent: 3,000 hours of professional project risk management experience
- If you have a high school diploma, associate's degree, or global equivalent: 4,500 hours of professional project risk management experience

**Contact Hours**

- If you have a bachelor's degree or global equivalent: 30 contact hours in the area of risk management
- If you have a high school diploma, associate's degree, or global equivalent: 40 contact hours in the area of risk management

**PMI Scheduling Professional (PMI-SP®)** The PMI-SP credential is ideal for those who specialize in the area of project scheduling, or who would like to highlight their project scheduling experience. You must meet the following requirements to apply for the PMI-SP exam:

**Work Experience** The following work experience must have been accrued over the past five consecutive years:

- If you have a bachelor's degree or global equivalent: 3,500 hours of professional project scheduling experience
- If you have a high school diploma, associate's degree, or global equivalent: 5,000 hours of professional project scheduling experience

**Contact Hours**

- If you have a bachelor's degree or global equivalent: 30 contact hours in the area of project scheduling
- If you have a high school diploma, associate's degree, or global equivalent: 40 contact hours in the area of project scheduling

For the latest information regarding the PMI credentials and other exam information, you can visit PMI's website at `www.pmi.org`.

# PMP Exam Objectives

The PMP exam tests your knowledge of the competencies highlighted in the exam objectives. The following are the official PMP exam objectives, as specified by PMI.

## Initiating the Project

The following objectives make up the Initiating the Project performance domain and are covered in Chapter 2 of this book:

- Perform project assessment based upon available information and meetings with the sponsor, customer, and other subject matter experts, in order to evaluate the feasibility of new products or services within the given assumptions and/or constraints.
- Define the high-level scope of the project based on business and compliance requirements, in order to meet the customer's project expectations.
- Perform key stakeholder analysis using brainstorming, interviewing, and other data-gathering techniques, in order to ensure expectation alignment and gain support for the project.
- Identify and document high-level risks, assumptions, and constraints based on the current environment, historical data, and/or expert judgment, in order to identify project limitations and propose an implementation approach.
- Develop the project charter by further gathering and analyzing stakeholder requirements, in order to document project scope, milestones, and deliverables.
- Obtain approval of the project charter from the sponsor and customer (if required), in order to formalize the authority assigned to the project manager and gain commitment and acceptance for the project.

## Planning the Project

The following objectives make up the Planning the Project performance domain and are covered in Chapter 3 of this book:

- Assess detailed project requirements, constraints, and assumptions with stakeholders based on the project charter, lessons learned from previous projects, and the use of requirement-gathering techniques (e.g., planning sessions, brainstorming, focus groups), in order to establish the project deliverables.

- Create the work breakdown structure with the team by deconstructing the scope, in order to manage the scope of the project.
- Develop a budget plan based on the project scope using estimating techniques, in order to manage project cost.
- Develop a project schedule based on the project timeline, scope, and resource plan, in order to manage timely completion of the project.
- Develop a human resource management plan by defining the roles and responsibilities of the project team members in order to create an effective project organization structure and provide guidance regarding how resources will be utilized and managed.
- Develop a communication plan based on the project organization structure and external stakeholder requirements, in order to manage the flow of project information.
- Develop a procurement plan based on the project scope and schedule, in order to ensure that required project resources will be available.
- Develop a quality management plan based on the project scope and requirements, in order to prevent the occurrence of defects and reduce the cost of quality.
- Develop a change management plan by defining how changes will be handled, in order to track and manage changes.
- Develop a risk management plan by identifying, analyzing, and prioritizing project risks and defining risk response strategies, in order to manage uncertainty throughout the project life cycle.
- Present the project plan to the key stakeholders (if required), in order to obtain approval to execute the project.
- Conduct a kickoff meeting with all key stakeholders, in order to announce the start of the project, communicate the project milestones, and share other relevant information.

## Executing the Project

The following objectives make up the Executing the Project performance domain and are covered in Chapter 4 of this book:

- Obtain and manage project resources, including outsourced deliverables, by following the procurement plan, in order to ensure successful project execution.
- Execute the tasks as defined in the project plan, in order to achieve the project deliverables within budget and schedule.
- Implement the quality management plan using the appropriate tools and techniques, in order to ensure that work is being performed according to required quality standards.
- Implement approved changes according to the change management plan, in order to meet project requirements.
- Implement approved actions (e.g., workarounds) by following the risk management plan, in order to minimize the impact of the risks on the project.
- Maximize team performance through leading, mentoring, training, and motivating team members.

## Monitoring and Controlling the Project

The following objectives make up the Monitoring and Controlling the Project performance domain and are covered in Chapter 5 of this book:

- Measure project performance using appropriate tools and techniques, in order to identify and quantify any variances, perform approved corrective actions, and communicate with relevant stakeholders.
- Manage changes to the project scope, schedule, and costs by updating the project plan and communicating approved changes to the team, in order to ensure that revised project goals are met.
- Ensure that project deliverables conform to the quality standards established in the quality management plan by using appropriate tools and techniques (e.g., testing, inspection, control charts), in order to satisfy customer requirements.
- Update the risk register and risk response plan by identifying any new risks, assessing old risks, and determining and implementing appropriate response strategies, in order to manage the impact of risks on the project.
- Assess corrective actions on the issue register and determine next steps for unresolved issues by using appropriate tools and techniques in order to minimize the impact on project schedule, cost, and resources.
- Communicate project status to stakeholders for their feedback, in order to ensure the project aligns with business needs.

## Closing the Project

The following objectives make up the Closing the Project performance domain and are covered in Chapter 6 of this book:

- Obtain final acceptance of the project deliverables by working with the sponsor and/or customer, in order to confirm that project scope and deliverables were met.
- Transfer the ownership of deliverables to the assigned stakeholders in accordance with the project plan, in order to facilitate project closure.
- Obtain financial, legal, and administrative closure using generally accepted practices, in order to communicate formal project closure and ensure no further liability.
- Distribute the final project report, including all project closure-related information, project variances, and any issues, in order to provide the final project status to all stakeholders.
- Collate lessons learned through comprehensive project review, in order to create and/or update the organization's knowledge base.
- Archive project documents and materials in order to retain organizational knowledge, comply with statutory requirements, and ensure availability of data for potential use in future projects and internal/external audits.
- Measure customer satisfaction at the end of the project by capturing customer feedback, in order to assist in project evaluation and enhance customer relationships.

# Project Foundation

Much of the focus and content of this book revolve heavily around the information contained in *A Guide to the Project Management Body of Knowledge, 4th Edition (PMBOK® Guide)*, published by the Project Management Institute (PMI). Because many exam questions will relate to the content of the *PMBOK® Guide*, it will be referenced throughout this book, and we'll elaborate further on those areas that appear on the test.

This chapter lays the foundation for building and managing a project. Understanding project management from a broad and high-level perspective is important and will prepare you to digest the rest of the information in this book.

# Defining a Project

Before delving into the aspects of project management, it's important to determine whether what you are dealing with is, in fact, a project. Projects are often confused with ongoing operations, and it's therefore important to understand how to define a project and know its characteristics.

Once you have determined that you are dealing with a project, all stakeholders will need to be identified. To be considered successful, a project must achieve its objectives and meet or exceed the expectations of the stakeholders.

For more detailed information on projects and stakeholders, see Chapter 1, "What Is a Project?" in *PMP: Project Management Professional Exam Study Guide, 6th Edition* (Sybex, 2011).

## Project Characteristics

The characteristics of projects, as shown in Figure 1.1, are as follows:

- They are temporary in nature and have definite start and end dates.
- They produce a unique product, service, or result that didn't exist before.

**FIGURE 1.1** Project

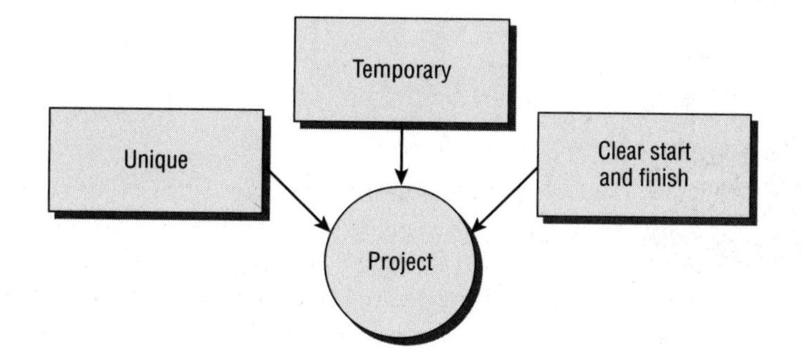


These products or services might include tangible products, services such as consulting or project management, and business functions that support the organization. Projects might also produce a result or an outcome, such as a document that details the findings of a research study. The purpose of a project is to meet its goals and to conclude. Therefore, a project is considered complete when one of the following occurs:

- The goals and objectives are accomplished to the satisfaction of the stakeholders.
- It has been determined that the goals and objectives cannot be accomplished.
- The project is canceled.

After its completion, a project's product, service, or result may become part of an ongoing operation.

## Operations

Operations are ongoing and repetitive, involving work that is continuous and without an end date. Often, operations involve repeating the same processes and producing the same results. Figure 1.2 shows the characteristics of operations. The purpose of operations is to keep the organization functioning.

**FIGURE 1.2** Operations

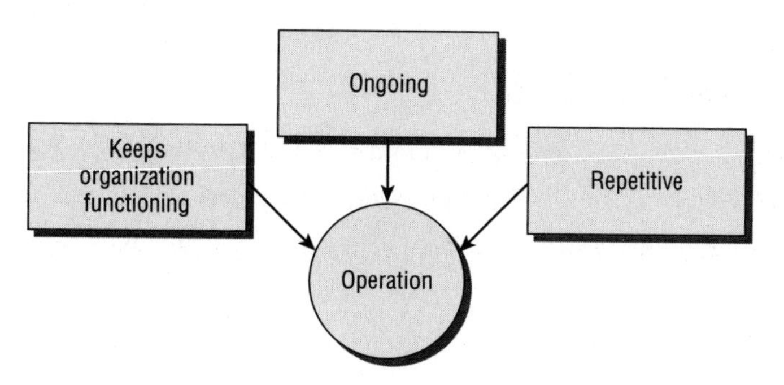

Unsure whether it is a project or ongoing operation? Put it through the following test to make a determination:

- Is it unique?
- Does the project have a limited time frame?
- Is there a way to determine when the project is completed?
- Is there a way to determine stakeholder satisfaction?

If you answered yes to each question, then it is a project.

## Progressive Elaboration

Progressive elaboration means the characteristics of the product, service, or result of the project are determined incrementally and are continually refined and worked out in detail as the project progresses. Product characteristics typically start out broad-based at the beginning of the project and are iterated into more and more detail over time until they are complete and finalized.

## Stakeholders

A stakeholder is any person or organization that is impacted by the project or who can impact the project. The following are characteristics of stakeholders:

- They are individuals or organizations with a vested interest in the project.
- They are actively involved with the work of the project.
- They have something to either gain or lose as a result of the project.

Identifying who these stakeholders are is not a onetime process, and it's important to identify stakeholders at the onset of the project. Here are some examples of project stakeholders, as shown in Figure 1.3:

- Customer
- Sponsor
- Contractors
- Suppliers
- Project manager
- Project team members
- Department managers

**FIGURE 1.3** Stakeholders

It's also important to note that stakeholders may have conflicting interests and that it is the project manager's responsibility to manage stakeholder expectations. And when in doubt, stakeholder conflicts should always be resolved in favor of the customer.

The project sponsor, who is also a stakeholder, is generally an executive in the organization with the authority to assign resources and enforce decisions regarding the project. The project sponsor typically serves as the tiebreaker decision maker and is one of the people on the project's escalation path.

Stakeholders are identified early on within the life of the project. Failure to identify stakeholders early may result in failure of the project itself. Understanding the level of influence of each stakeholder is also critical to the success of the project.

## Exam Essentials

**Be able to describe the difference between projects and operations.** A project is temporary in nature with a definite beginning and ending date. Projects produce unique products, services, or results. Operations are ongoing and use repetitive processes that typically produce the same result over and over.

# Defining Project Management

Project management provides the tools and techniques necessary for the successful initiation, planning, and execution of a project. Project management may involve more than a single project. In the bigger picture, a project may be part of a program, a portfolio, and/or a project management office.

For more detailed information on project management basics, see Chapter 1 of *PMP: Project Management Professional Exam Study Guide, 6th Edition.*

## Project Management

According to the *PMBOK® Guide,* project management is "the application of knowledge, skills, tools, and techniques to project activities to meet the project requirements."

Project managers are the people responsible for managing the project processes and applying the tools and techniques used to carry out the project activities. It is the responsibility of the project manager to ensure that project management techniques are applied and followed. In addition to this, project management is a process that includes initiating a new project, planning, putting the project plan into action, and measuring progress and performance. It involves identifying the project requirements, establishing project objectives, balancing constraints, and taking the needs and expectations of the key stakeholders into consideration.

Project management itself can exist beyond the management of a single project. In some organizations, programs and portfolios are also managed. Figure 1.4 shows the characteristics of a project, program, portfolio, as well as a project management office (PMO).

**FIGURE 1.4** Project management overview

| Project Management | | | |
|---|---|---|---|
| Project | Program | Portfolio | PMO |
| • Produces a unique product, service, or result | • Groups of related projects<br>• Managed using similar techniques | • Collections of programs and projects<br>• Supports specific business goals | • Centralized organizational unit<br>• Oversees projects and programs |

## Programs

According to the *PMBOK® Guide*, programs are groups of related projects that are managed using the same techniques in a coordinated fashion, allowing them to capitalize on benefits that wouldn't be achievable otherwise.

Each subproject within a program has its own project manager, who reports to the program manager. All the projects are related and are managed together so that collective benefits are realized and controls are implemented and managed in a coordinated fashion. The management of this collection of projects—determining their interdependencies, managing their constraints, and resolving issues among them—is called program management.

## Portfolios

Portfolios are collections of programs and projects that support a specific business goal or objective. Programs and projects included as part of a portfolio may not necessarily relate to one another in a direct way.

Portfolio management encompasses managing the collection of programs, projects, other work, and sometimes other portfolios. It also concerns monitoring active projects for adherence to objectives, balancing the portfolio among the other investments of the organization, and assuring the efficient use of resources.

## Project Management Office

The project management office (PMO) is a centralized organizational unit that oversees the management of projects and programs throughout the organization.

The PMO is responsible for the following:

- Managing the objectives of a collective set of projects
- Managing resources across the projects
- Managing the interdependencies of all the projects within its authority
- Maintaining and archiving project documentation for future reference
- Measuring project performance of active projects and suggesting corrective actions
- Evaluating completed projects for their adherence to the project plan
- Suggesting corrective actions pertaining to projects and programs as needed

The most common reason a company starts a project management office is to establish and maintain procedures and standards for project management methodologies and to manage resources assigned to the projects in the PMO. Project managers and team members may report directly to the PMO, or the PMO may provide support functions for projects and project management training or simply have experts available to provide assistance.

The overall purpose of a project manager and PMO differ to a certain extent. A project manager is focused on the project at hand, while the PMO takes more of an organizational approach to project management. While the project manager is concerned with the project's objectives, the PMO is concerned with the organizational objectives. Table 1.1 outlines the differences between the overall objectives and perspectives of a project manager and PMO.

**TABLE 1.1** Project manager vs. PMO

| | Project Manager | PMO |
|---|---|---|
| **Overall goal** | Achieve the project requirements | Achieve goals from an enterprise-wide perspective |
| **Objective** | Deliver project objectives | Deliver organizational objectives |
| **Resources** | Manage project resources | Optimize shared resources |
| **Reporting** | Project reporting | Consolidated reporting |

PMOs can exist in any type of organizational structure. A major benefit for project managers working in an organization that has an existing PMO is the room for advancement that a PMO provides.

While PMOs are common in organizations today, they are not required in order to apply good project management practices.

**Exam Essentials**

**Be able to differentiate between project management, program management, portfolio management, and a project management office.** Project management brings together a set of tools and techniques to describe, organize, and monitor the work of project activities; program management refers to the management of groups of related projects; portfolio management refers to the management of a collection of programs and projects that support a specific business goal or objective; and the project management office oversees the management of projects and programs throughout the organization.

# Identifying Project Management Skills

To be successful, a project manager must possess general management skills that span every area of management—from accounting and strategic planning to supervision, personnel administration, and so forth. Some projects also require specific skills in certain application areas, such as within an industry group, within a department, or by technical specialty. These skills prepare the project manager to communicate, solve problems, lead, and negotiate through a project.

For more detailed information on project management skills, of *PMP: Project Management Professional Exam Study Guide,*

As shown in Figure 1.5, seven general management skills make up the foundation of good project management practices.

**FIGURE 1.5** General management skills

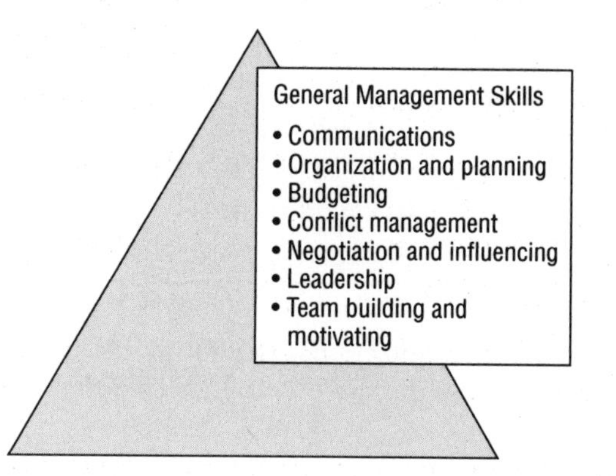


They are as follows:

**Communication Skills** Many forms of communication exist within a project and are critical to its success. It is the job of the project manager to ensure that the information is explicit, clear, and complete. Once the information has been distributed, it is the responsibility of the recipient to confirm that the information has been understood. *A good project manager spends up to 90 percent of their time communicating.* This makes communication skills a critical asset for a project manager.

**Organization and Planning Skills** Organization and planning skills are closely related and an important skill set that a project manager should possess. Project managers must track and locate many documents, meetings, contracts, and schedules. Doing so requires time management skills, which is why this skill set is tied closely to organizational skills. Because there isn't any aspect of project management that doesn't first involve planning, it will be an area discussed extensively throughout this book.

**Budgeting Skills** Project managers establish and manage budgets and are therefore expected to have some knowledge of finance and accounting principles. Cost estimates are an example of when budgeting skills are put to use. Other examples include project spending, reading and understanding vendor quotes, preparing or overseeing purchase orders, and reconciling invoices.

**Conflict Management Skills** Conflict management involves problem solving, first by defining the causes of the problem and then by making a decision to solve it. After examining and analyzing the problem, the situation causing it, and the available alternatives, determine the best course of action. The timing of the decision is also important.

Both negotiation and influencing skills are necessary ... utilized in all areas of project management. The fol- ... ll set:

one or with teams of people, negotiation involves working ... eement and is necessary in almost every area of the project.

... quires an understanding of the organization's structure and ... er party to consider the choice you think is the better one ... want.

... ique used to influence people and is the ability to get people to ... otherwise do.

... ves getting groups of people with different interests to cooperate creatively ... midst of conflict and disorder.

**Leadership Skills** Leadership involves exhibiting characteristics of both a leader and a manager and knowing when to switch from one to the other throughout the project. A leader imparts vision, gains consensus for strategic goals, establishes direction, and inspires and motivates others. Managers focus on results and are concerned with getting the job done according to requirements.

**Team Building and Motivating Skills** The project manager sets the tone for the project team and walks the team members through various stages of team development to become fully functional. This can involve team-building groundwork and motivating the team even when team members are not direct reports.

**Exam Essentials**

**Be able to list some of the skills every good project manager should possess.** Communication, budgeting, organization, problem solving, negotiation and influencing, leading, and team building are skills a project manager should possess.

# Understanding Organizational Structures

Like projects, organizations are unique, each with its own style and culture, which both have an influence on how projects are performed. Organizations in general can be structured in one of three ways:

- Functional
- Matrix
- Projectized

For more detailed information on project management skills, see Chapter 1 of *PMP: Project Management Professional Exam Study Guide, 6th Edition.*

As shown in Figure 1.5, seven general management skills make up the foundation of good project management practices.

**FIGURE 1.5** General management skills

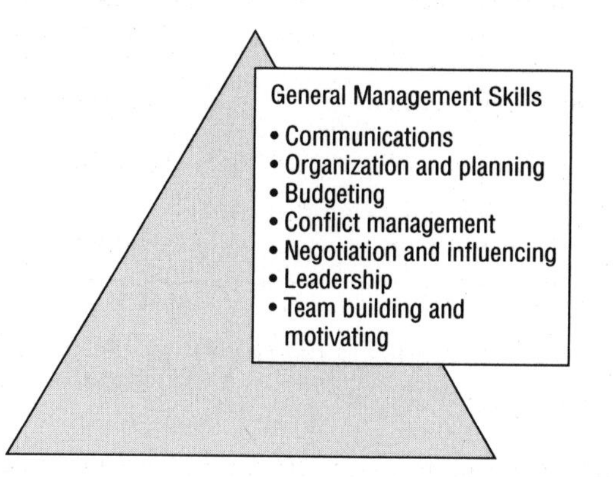


They are as follows:

**Communication Skills** Many forms of communication exist within a project and are critical to its success. It is the job of the project manager to ensure that the information is explicit, clear, and complete. Once the information has been distributed, it is the responsibility of the recipient to confirm that the information has been understood. *A good project manager spends up to 90 percent of their time communicating.* This makes communication skills a critical asset for a project manager.

**Organization and Planning Skills** Organization and planning skills are closely related and an important skill set that a project manager should possess. Project managers must track and locate many documents, meetings, contracts, and schedules. Doing so requires time management skills, which is why this skill set is tied closely to organizational skills. Because there isn't any aspect of project management that doesn't first involve planning, it will be an area discussed extensively throughout this book.

**Budgeting Skills** Project managers establish and manage budgets and are therefore expected to have some knowledge of finance and accounting principles. Cost estimates are an example of when budgeting skills are put to use. Other examples include project spending, reading and understanding vendor quotes, preparing or overseeing purchase orders, and reconciling invoices.

**Conflict Management Skills** Conflict management involves problem solving, first by defining the causes of the problem and then by making a decision to solve it. After examining and analyzing the problem, the situation causing it, and the available alternatives, determine the best course of action. The timing of the decision is also important.

**Negotiation and Influencing Skills** Both negotiation and influencing skills are necessary in effective problem solving and are utilized in all areas of project management. The following terms are related to this skill set:

**Negotiation** Whether one on one or with teams of people, negotiation involves working with others to come to an agreement and is necessary in almost every area of the project.

**Influencing** Influencing requires an understanding of the organization's structure and involves convincing the other party to consider the choice you think is the better one even if it is not what they want.

**Power** Power is a technique used to influence people and is the ability to get people to do things they wouldn't otherwise do.

**Politics** Politics involves getting groups of people with different interests to cooperate creatively even in the midst of conflict and disorder.

**Leadership Skills** Leadership involves exhibiting characteristics of both a leader and a manager and knowing when to switch from one to the other throughout the project. A leader imparts vision, gains consensus for strategic goals, establishes direction, and inspires and motivates others. Managers focus on results and are concerned with getting the job done according to requirements.

**Team Building and Motivating Skills** The project manager sets the tone for the project team and walks the team members through various stages of team development to become fully functional. This can involve team-building groundwork and motivating the team even when team members are not direct reports.

**Exam Essentials**

**Be able to list some of the skills every good project manager should possess.** Communication, budgeting, organization, problem solving, negotiation and influencing, leading, and team building are skills a project manager should possess.

# Understanding Organizational Structures

Like projects, organizations are unique, each with its own style and culture, which both have an influence on how projects are performed. Organizations in general can be structured in one of three ways:

- Functional
- Matrix
- Projectized

Within these structures, variations and combinations exist. Figure 1.6 shows where these structures fall in relation to one another, including the variations that exist in matrix organizations. Knowing the type of structure that an organization operates within is important and impacts how a project manager may proceed. The organizational structure is part of the project environment, which influences how the project is managed. One of the biggest determining factors of an organization's structure is the authority that a project manager holds.

**FIGURE 1.6** Spectrum of organizational types

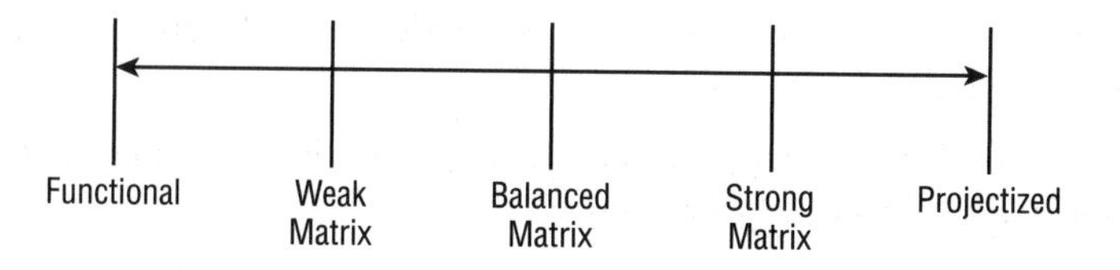


Since functional and projectized are the two extreme types of organizational structures, we'll cover them first.

For more detailed information on functional, projectized, and matrix organizations, including sample charts, see Chapter 1 of *PMP: Project Management Professional Exam Study Guide, 6th Edition.*

## Functional Organizations

Functional organizations are the oldest style of organization; the functional style is known as the traditional approach to organizing businesses. These types of organizations have the following characteristics:

- They center around specialties.
- They are grouped by function.
- They are displayed as a hierarchy.

Importantly, workers within a functional organization specialize in an area of expertise, allowing them to become very good at their specialty.

Within this hierarchical structure, a chain of command exists in which employees report to one manager and one person is ultimately in charge at the top. Each department or group is managed independently and has a limited span of control, therefore requiring that the chain of command be followed when input from another department on a project is needed. The mind-set behind this type of structure is that people with similar skills and experiences are easier to manage as a group, allowing for work assignments to be distributed and managed easily, and supervisors are experienced in the area they supervise.

Table 1.2 highlights the advantages and disadvantages of a functional organization.

**TABLE 1.2** Functional organizations

| Advantages | Disadvantages |
|---|---|
| There is an enduring organizational structure. | Project managers have little to no formal authority. |
| There is a clear career path with separation of functions, allowing specialty skills to flourish. | Multiple projects compete for limited resources and priority. |
| Employees have one supervisor with a clear chain of command. | Project team members are loyal to the functional manager. |

## Projectized Organizations

Projectized organizations are nearly the opposite of functional organizations, focusing on the project itself and therefore developing loyalties to the project and not to a functional manager.

Project managers have high to ultimate authority within this type of organization and typically report directly to the CEO. The project manager makes decisions regarding the project, including dealing with constraints. As a side note, project managers in all organizational structures must balance competing project constraints such as these:

- Scope
- Quality
- Schedule
- Budget
- Resources
- Risk

Within a projectized organization, teams are often co-located, meaning that team members physically work at the same location and report to the project manager. Once the project is completed, the team is often dissolved and members are either put on the bench or let go. Resource utilization can be inefficient if a highly specialized skill is needed only at given times throughout the project and sits idle during other times.

Table 1.3 highlights the advantages and disadvantages of a projectized organization.

**TABLE 1.3** Projectized organizations

| Advantages | Disadvantages |
|---|---|
| Project managers have high to ultimate authority. | Project team members may find themselves out of work after project completion. |
| Project team members are loyal to the project. | There can be inefficiency with resource utilization. |

## Matrix Organizations

Matrix organizations are a blend of functional and projectized organizations, taking advantage of the strengths of both and combining them into one. Employees report to a functional manager and to at least one project manager.

Functional managers assign employees to projects and handle administrative duties, while project managers execute the project and assign project activities to employees. Both share the responsibility of performance reviews.

Within matrix organizations exists a balance of power ranging from that of a functional organization to that of a projectized organization:

**Strong Matrix** A strong matrix organization has characteristics of a projectized organization, where the project manager has a high level of authority and can dictate resources.

**Weak Matrix** On the other end of the spectrum is the weak matrix, which holds a resemblance to a functional organization in that the functional managers have a greater level of power and tend to dictate work assignments.

**Balanced Matrix** In between a strong matrix and weak matrix is the balanced matrix, where the power is balanced between the project managers and functional managers.

Table 1.4 shows the subtle differences between the three types of matrix organizations. The majority of organizations today are a composite of functional, projectized, and matrix structures.

**TABLE 1.4** Comparing matrix structures

| | Weak Matrix | Balanced Matrix | Strong Matrix |
|---|---|---|---|
| Project manager's title | Project coordinator, project leader, or project expeditor | Project manager | Project manager |
| Project manager's focus | Split focus between project and functional responsibilities | Projects and project work | Projects and project work |
| Project manager's power | Minimal authority and power | Balance of authority and power | Significant authority and power |
| Project manager's time | Part-time on projects | Full-time on projects | Full-time on projects |
| Organization style | Most like functional organization | Blend of both weak and strong matrix | Most like a projectized organization |
| Project manager reports to | Functional manager | Functional manager, but shares authority and power | Manager of project managers |

**Exam Essentials**

**Be able to differentiate between the organizational structures and the project manager's authority in each.** Organizations are usually structured in some combination of the following: functional, projectized, and matrix (including weak matrix, balanced matrix, and strong matrix). Project managers have the most authority in a projectized organization and the least authority in a functional organization.

# Understanding the Project Environment

The project environment is made up of internal and external factors that influence a project. When managing a project, the project manager must consider more than just the project itself. Proactively managing a project involves understanding the environment in which the project must function.

The following are some examples of the elements that make up the project environment:

**Organizational Structure** The organizational structure itself plays a large role in how a project is managed and is a major influential factor within the project environment. This is intermingled with the company culture.

**Physical Environment** The physical environment includes considerations such as the following:

- Local ecology
- Physical geography
- Environmental restrictions, such as protected areas or protected wildlife

**Cultural and Social Environment** When referencing the cultural environment, there are various layers to consider:

- Corporate culture, including external stakeholders and vendors. This is intermingled with organizational structures but can take on a life of its own.
- Ethnic culture.
- Religious culture.

To communicate effectively with customers, stakeholders, and project team members, it becomes necessary to understand cultural and social factors and how they may impact a project. In particular, cultural differences pose a common communication obstacle within projects. As virtual and global projects become a more common scenario, understanding how to function within this type of environment becomes critical.

A project manager should also understand how the project impacts individuals and how individuals impact the project.

**International and Political Environment** The international and political environment involves considering the following:

- International, national, regional, and local laws and regulations
- Customs
- Political climate
- Time zone and local holidays

It is important for a project manager to know, from the onset of the project, whether any of the items listed affect their particular project. Something as minor as a time zone difference or as large as the political climate can have a huge impact on a project.

There are other influential factors far beyond those listed here, and the list can become quite extensive. On a basic level, these factors provide you with an overview of a project from a broader perspective.

# Understanding Project Life Cycles and Project Management Processes

Most projects begin with an idea. After soliciting support and obtaining approval, a project progresses through the intermediate phases to the ending phase, where it is completed and closed out. As a project passes through this life cycle, it is carried out through a set of project management processes. According to the *PMBOK® Guide*, these processes are interrelated and dependent on one another; they are divided into five project management process groups:

- Initiating
- Planning
- Executing
- Monitoring and Controlling
- Closing

Each of these process groups displays characteristics that reflect the level of a project's costs, staffing, chances for successful completion, stakeholder influence, and probability of risk.

Figure 1.7 shows the project management process groups and represents the life cycle of a project or project phase.

**FIGURE 1.7** Project management process groups

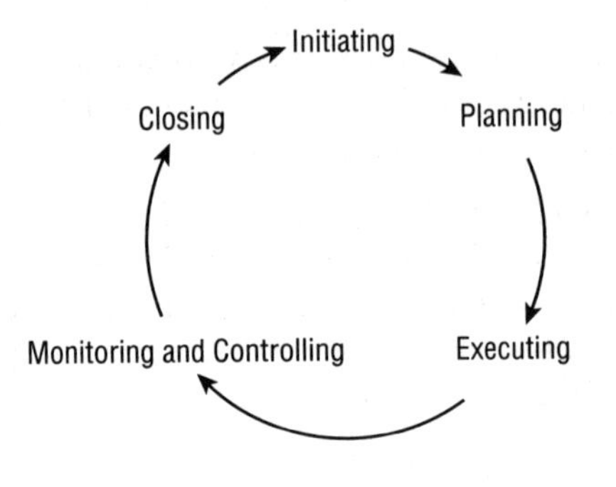

For more detailed information on the project management life cycle and processes, see Chapter 1 of *PMP: Project Management Professional Exam Study Guide, 6th Edition.*

## Project Phases and Project Life Cycles

Most projects are divided into phases. Project phases consist of segments of work, and a project may have one or many phases, depending on the project complexity and industry. The phases that a project progresses through are collectively called the *project life cycle.* Project life cycles are similar for all projects, and the phases that occur within the project life cycle are sequential and sometimes overlap.

According to the *PMBOK® Guide,* all projects follow the same life-cycle structure:

- Initiating the project
- Planning the work of the project
- Performing the work of the project
- Closing out the project

At the beginning of a phase, a feasibility study may be carried out, and at the end of a phase, a phase-end review of the accomplished deliverables may be performed before hand-off to the next phase can occur. The following list expands further on several key terms used within this description of what occurs within a phase:

**Handoffs** For the project to progress from one phase to the next, the phase deliverables must be reviewed for accuracy and approved. As each phase is completed, it's handed off to the next phase. Handoffs, or technical transfers, are phase sequences that signal the end of one phase and typically mark the beginning of the next.

**Feasibility Studies** Some projects incorporate feasibility studies in the beginning phase. They are completed prior to the beginning of the next phase. A feasibility study determines whether the project is worth undertaking and whether it will be profitable to the organization.

**Deliverables** Each phase has at least one deliverable that marks the phase completion. A deliverable is a tangible output that can be measured and must be produced, reviewed, and approved to bring the phase or project to completion.

**Phase-End Reviews** At the end of a phase, a phase-end review takes place so that those involved with the work may determine whether the project should continue on to the next phase. Phase-end reviews, also known as phase exits, phase gates, milestones, stage gates, decision gates, and kill points, give the project manager the ability to identify and address errors discovered during the phase.

**Phase-to-Phase Relationships** As previously indicated, a project may consist of one or more phases. The *PMBOK® Guide* defines three types of phase-to-phase relationships:

- Sequential relationships, where one phase must finish before the next phase can begin
- Overlapping relationships, where one phase starts before the prior phase completes
- Iterative relationships, where work for subsequent phases is planned as the work of the previous phase is performed

The act of overlapping phases to shorten or compress the project schedule is called fast-tracking. This means that a later phase is started prior to completing and approving the phase or phases that come before it.

## Project Management Process Groups

Project management processes organize and describe the work of the project. As mentioned earlier, the following are the five process groups described in the *PMBOK® Guide*:

- Initiating process group
- Planning process group
- Executing process group
- Monitoring and Controlling process group
- Closing process group

These processes are performed by people; they are interrelated and dependent on one another and include individual processes that collectively make up the group. The five process groups are iterative and might be revisited and revised throughout the project's life as the project is refined.

The project manager and the project team are responsible for tailoring, which involves determining which processes within each process group are appropriate for the project. When the project is tailored, its size and complexity and various inputs and outputs are taken into consideration. The conclusion of each process group allows the project manager and stakeholders to reexamine the business needs of the project and determine whether the project is satisfying those needs.

**Initiating Process Group** The processes within the Initiating process group occur at the beginning of the project and at the beginning of each project phase for large projects. This process group grants the approval to commit the organization's resources to working on the project or phase and authorizes the project manager to begin working on the project.

According to the *PMBOK® Guide*, the Initiating process group contains the following actions:

- Creating the project charter, which formally authorizes the project
- Identifying the project stakeholders

**Planning Process Group** The Planning process group includes processes that formulate and revises project goals and objectives. It is also where the project management plan that will be used to achieve the goals the project was undertaken to address is created. Within this process group, project requirements are fleshed out, and stakeholders are identified. Because projects are unique, planning must encompass all areas of project management.

According to the *PMBOK® Guide*, the Planning process group contains the following actions:

- Creating the project management plan

  The project management plan itself contains several subsidiary plans and baselines. This collection of plans defines how the project will be planned, carried out, and closed.
- Collecting project requirements

  The project requirements define and document the needs of the stakeholders.
- Developing the project and product's scope
- Creating the work breakdown structure (WBS)

  The WBS contains a decomposition of the project work, from the major deliverables down to the work packages.
- Identifying the project activities necessary to carry out the project work
- Developing the order and duration of project activities
- Calculating the number of resources needed to carry out the project activities
- Creating the project schedule
- Developing the project's budget
- Estimating the cost of each project activity
- Planning and creating the quality management plan

  The quality management plan sets the standards for the project and also defines the quality requirements.
- Creating the human resource plan

  This plan outlines how human resources will be identified and managed.

- Creating the communications management plan

  This plan defines how communications will be carried out throughout the course of the project, including identifying and meeting the communication needs of stakeholders.

- Creating the risk management plan

  This plan outlines how risk management will be conducted. It also provides the necessary information to consistently identify and measure project risk.

- Identifying project risks

  This process also results in the creation of the risk register, where information about the identified risks will be documented.

- Assessing and prioritizing project risks that require further analysis or action, as well as the numerical analysis of risk on the overall project objectives
- Developing risk responses
- Creating the procurement management plan

  The procurement management plan documents and describes how external products or services will be acquired.

**Executing Process Group** The Executing process group involves putting the project management plan into action, keeping the project plan on track, and implementing approved changes.

The project manager coordinates and directs project resources, and costs are highest here since greater amounts of time and resources are utilized during the Executing processes.

According to the *PMBOK® Guide*, the Executing process group contains the following actions:

- Performing the work defined in the project management plan
- Ensuring that the quality standards and operational definitions outlined in the planning process group are followed
- Auditing the quality requirements
- Obtaining the human resources needed to complete the project work
- Enhancing the skills and interactions of the project team as well as the overall team environment
- Managing the project team and enhancing its performance
- Distributing the necessary information throughout the course of the project, as outlined within the communications management plan
- Managing the expectations of the stakeholders
- Selecting the external vendors used and awarding the necessary procurement contracts

**Monitoring and Controlling Process Group** The Monitoring and Controlling process group is where project performance measurements are taken and analyzed to determine whether the project is staying in line with the project plan. This process group tracks the progress of work being performed and identifies problems and variances within a process group and the project as a whole.

According to the *PMBOK® Guide*, processes from within the Monitoring and Controlling process group contain the following actions:

- Monitoring and controlling the work of the project, such as making sure the actual work performed is in line with the project management plan
- Using the integrated change control system to review, approve or deny, and manage all changes within the project
- Accepting the completed project deliverables
- Controlling the project and product scope

  This helps prevent scope creep and makes sure the project accomplishes what it set out to do.
- Controlling the project schedule to make sure the project is completed on time
- Controlling the project costs to make sure the project is completed within the allotted budget
- Monitoring the results of conducting quality activities

  This allows for an assessment of performance.
- Reporting the project performance through planned reports

  This also involves forecasting and measuring project performance.
- Implementing risk response plans
- Monitoring and controlling risks

  This includes monitoring the project for any new risks and evaluating the effectiveness of risk responses and risk management processes.
- Managing vendor relationships and vendor contracts

**Closing Process Group** The Closing process group includes processes that bring a formal, orderly end to the activities of a project phase or to the project itself. Once the project objectives have been met, all the project information is gathered and stored for future reference. Contract closeout also occurs within this process group, and formal acceptance and approval are obtained from project stakeholders.

According to the *PMBOK® Guide*, the closing process group contains the following actions:

- Finalizing the project work to formally complete the project or project phase
- Completing and then closing out all vendor contracts

The progression through the project management process groups exhibits the same characteristics as progression through the project phases. Table 1.5 shows the characteristics of the process groups.

**TABLE 1.5** Characteristics of the project process groups

| | Initiating | Planning | Executing | Monitoring and Controlling | Closing |
|---|---|---|---|---|---|
| Costs | Low | Low | Highest | Lower | Lowest |
| Staffing levels | Low | Lower | High | High | Low |
| Chance for successful completion | Lowest | Low | Medium | High | Highest |
| Stakeholder influence | Highest | High | Medium | Low | Lowest |
| Risk probability | Highest | High | Medium | Low | Lowest |

## Plan-Do-Check-Act Cycle

As the project progresses and more information becomes known, the project management processes might be revisited and revised to update the project management plan. Underlying the concept that process groups are iterative is a cycle called the Plan-Do-Check-Act cycle, which was originally defined by Walter Shewhart and later modified by Edward Deming. The idea behind this concept is that each element in the cycle is results-oriented and becomes input into the next cycle. The cycle interactions can be mapped to work in conjunction with the five project management process groups.

**Exam Essentials**

**Be able to name the five project management process groups.** The five project management process groups are Initiating, Planning, Executing, Monitoring and Controlling, and Closing.

# Recognizing Professional and Social Responsibility

A PMP credential comes with several responsibilities that involve decisions and actions you make. When submitting your PMP application, you are agreeing to abide by the *PMI Code of Ethics and Professional Conduct*. This oath touches on several elements within your personal and professional life and includes actions that are mandatory as well as actions and ideals to which we aspire.

Maintaining ethical and professional conduct applies to day-to-day activities and is therefore applicable to everything discussed earlier within this chapter and throughout the rest of the book. Meeting ethical and professional standards is an essential part of building a solid project management foundation. For the exam, you will be expected to be familiar with the code and answer questions based on its contents.

# Upholding Individual Integrity

The *PMI Code of Ethics and Professional Conduct* addresses the following four elements: responsibility, respect, fairness, and honesty.

## Responsibility

Responsibility is a major part of upholding integrity. It involves taking ownership for your actions, being proactive in doing the right thing, and doing what is necessary to resolve a situation that results from an initial lack of responsibility.

As shown in Figure 1.8, there are a variety of elements that touch on the concept of responsibility:

- Ensuring a high level of integrity
- Accepting only those assignments for which you are qualified
- Following rules, regulations, laws, and policies
- Respecting confidential information

**FIGURE 1.8** Elements of responsibility

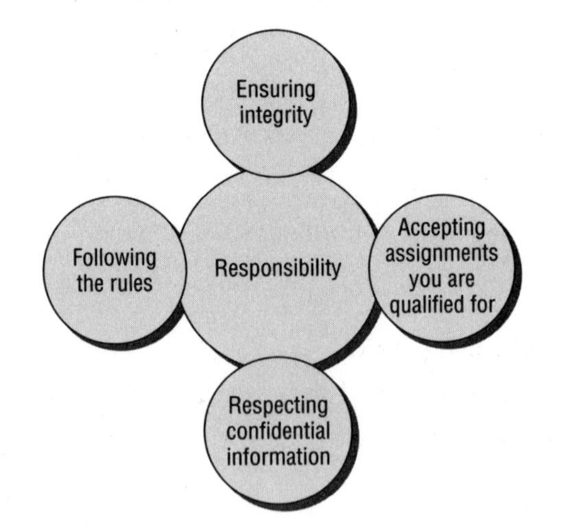

Responsibility involves making decisions that are for the good of the organization rather than for yourself. It also involves admitting your mistakes and being accountable for the decisions you make and the consequences that result from those decisions and actions.

### Ensuring Integrity

As a project manager, one of your professional responsibilities is to ensure integrity of the project management processes, the product, and your own personal conduct. This can entail the following:

- Applying the project management processes correctly to ensure integrity of the product
- Executing the process groups effectively to ensure the quality of the product
- Obtaining formal acceptance of the project and product

### Accepting Assignments

Accepting appropriate assignments is tied to responsibility and honesty. This involves doing the following:

- Accurately and truthfully reporting your qualifications, experience, and past performance of services to potential employers, customers, PMI, and others
- Declining assignments beyond your capabilities or experience
- Emphasizing your knowledge and how you've used it in your specific industry as opposed to making others believe that you have experience using certain processes and tools if, in reality, you don't
- Accurately representing yourself, your qualifications, and your estimates in your advertising and in person

### Following the Rules

It is imperative that you follow all applicable laws, rules, and regulations, including the following:

- Industry regulations
- Organization rules
- Project rules
- PMI organizational rules and policies
- Intellectual property and other ethical standards and principles governing your industry
- Local, national, and international law governing your location, processes, and product

Rules or regulations that apply in one country may or may not apply in other countries.

### Respecting Confidential Information

Whether you are a direct employee, a consultant, or a contractor, it is likely that at some point, you will come across some form of sensitive information. You may collect sensitive information as part of a project. You may be required to provide such information to a contractor. You may have access to this type of information as part of your role. Whatever the situation, it is important that you comply with the *PMI Code of Ethics and Professional Conduct,* which states that you agree not to disclose sensitive or confidential information or use it in any way for personal gain.

Handling company data, trade secrets, data from governmental agencies, and intellectual property are typical cases in which you will come across sensitive or confidential information. Don't forget to ask yourself the key questions shown in Figure 1.9 whenever you handle information. If you are unsure of the answer to any of the questions, find out before you share the information.

**FIGURE 1.9** Handling information

| Ask Yourself: | | |
|---|---|---|
| Is this information private or protected? | Can this information be shared? | Does a nondisclosure agreement exist? |

**Trade Secrets** An organization's trade secrets or private information should not be used for personal gain. Understanding what information may be shared with vendors, other organizations, or even coworkers is important. In practice, there are often gray areas where the breach of information is not as obvious. Being conscious and sensitive about your organization's information can make a big difference. If ever in doubt, always ask.

**Restricted Access** Working with the government is a great example of when data sharing can be sensitive. One governmental agency does not always have access to data from another, separate governmental agency. Depending on the data, others may have restricted access or no access at all. Restricted access isn't limited to governmental data; similar restrictions occur in the private sector as well. One department within an organization may not always have access to data from another department. Think of the sensitive nature of information housed within the human resources department. An employee's file may not always be accessible to department heads or others. It is a mistake to assume that others should have access to data because you believe it is logical that they do.

**Nondisclosure Agreements** It is your responsibility to make sure vendors or organizations that will have access to sensitive data sign the appropriate nondisclosure agreements before you release the data.

### Intellectual Property

You are likely to come into contact with intellectual property during the course of your project management career. The business or person who created the protected material or item owns the intellectual property. Intellectual property typically includes the following items:

- Items developed by an organization that have commercial value but are not tangible
- Copyrighted or trademarked material, such as books, software, and artistic works
- Ideas or processes that are patented
- An industrial process, a business process, or manufacturing process developed by the organization for a specific purpose

You might have to pay royalties or request written permission to use the property. Intellectual property should be protected just like sensitive or confidential data.

## Respect

Demonstrating respect involves how you conduct yourself, the way you treat others and the resources around you, and listening to the viewpoints of others. Respect is something that should be exhibited on a personal and professional level. Among other things, the level of respect you earn will be based on how well you do the following:

- Exhibit a professional demeanor
- Handle ethics violations
- Demonstrate a sense of cultural awareness
- Interpret experiences

### Exhibiting Professional Demeanor

Exhibiting a professional demeanor is an important part of business. You are responsible for your own actions and reactions. Here are some examples of maintaining a professional demeanor and encouraging professional demeanor within others:

- Controlling your emotions and your reactions in questionable situations
- Notifying the project team during the kickoff meeting where they can find a copy of organizational policies regarding conflict of interest, cultural diversity, standards and regulations, customer service, and standards of performance
- Coaching and influencing team members who exhibit nonprofessional attitudes or behaviors to conform to the standards of conduct expected by you and your organization. As the project manager, it is your job to ensure that team members act professionally because they represent you and the project.

### Reporting Ethics Violations

One of the responsibilities that falls to you as a PMP is the responsibility to report violations of the PMP code of conduct. To maintain integrity of the profession, PMP credential holders must adhere to the code of conduct that makes all of us accountable to one another. This includes notifying PMI when you know a violation has occurred and you've verified the facts. The *PMI Code of Ethics and Professional Conduct* includes the following violations:

- Conflicts of interest
- Untruthful advertising
- Falsely claiming PMP experience and credentials
- Appearances of impropriety

### Demonstrating Cultural Awareness

As project teams and organizations become increasingly more diverse, the need for a heightened awareness of cultural influences and customary practices becomes critical. Today, it's common for project teams to be spread out globally. It is rare *not* to come into contact with someone from another culture. For example, you may come across some or all of the following:

- Virtual project teams with team members based in different physical locations
- Internal project stakeholders spread among offices globally
- Offshore vendors or suppliers
- Offshore partners
- Project team members who, while physically located in the same office, are from different cultures

It is important not to attempt to force your own culture or customs on others. When in a different country, you will need to respect the local customs. At the same time, it is not acceptable to do something that violates your ethics and the ethics of your company.

As part of cultural awareness, you will need to be proactive in protecting yourself and your staff from experiencing culture shock. In order to do this, it becomes important to devote time to diversity training as needed, and to always demonstrate respect for your neighbors, as follows:

**Culture Shock** Working in a foreign country can bring about an experience called culture shock. Culture shock involves a separation from what you would normally expect. This can cause disorientation, frustration, breakdown of communication, and confusion. Here are some ways to prepare yourself and protect against culture shock (Figure 1.10):

- Read about the country you're going to work in before you go.
- Become familiar with customs and acceptable practices in foreign countries where you have dealings or will be visiting.
- When in doubt about a custom or situation, ask your hosts or a trusted contact from the company for guidance.

**FIGURE 1.10** Avoiding culture shock

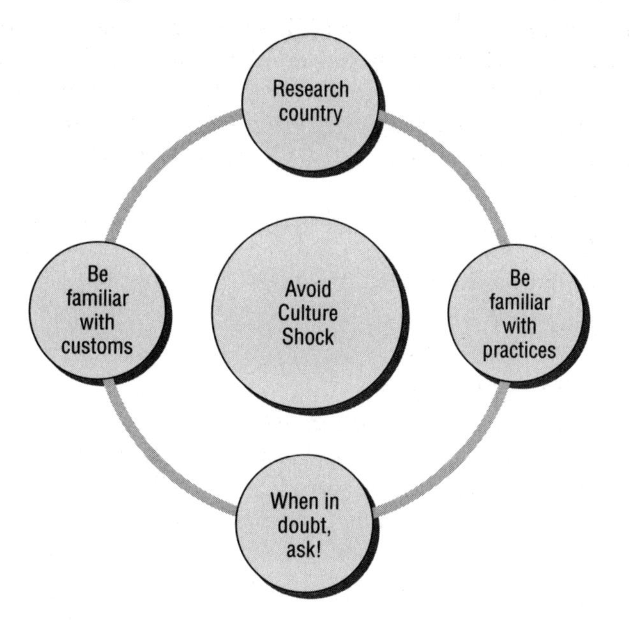


**Diversity Training** Project team members may be from different countries or cultures. Providing your team members with training helps to prevent cultural or ethical differences from impacting the project.

Diversity training makes people aware of differences between cultures and ethnic groups and helps them gain respect and trust. Training involving the company culture should also be provided so that project team members understand how the organization functions.

**Building Relationships** Some cultures require that a relationship be established before business can be conducted. Building that relationship may take days, depending on the culture.

Building relationships is valuable, regardless of whether or not a culture requires it. On the flip side, as you are building relationships, remember that you should not impose your beliefs on others because that is not the purpose of building relationships. Instead, building a relationship with others creates an atmosphere of mutual trust and cooperation. In this environment, all aspects of project planning and management, including negotiating and problem solving, become easier and smoother to navigate.

## Interpreting Experiences

As part of exhibiting respect, it's important to be conscious of how experiences are interpreted and perceived. This is relevant on both ends of the communication spectrum. Others may interpret your actions in a way that you did not intend. Likewise, you may misinterpret the actions of others. Remember the following:

- Each member of your team brings a unique set of experiences to the project. Those experiences influence perception, behavior, and decisions.
- Your perception of a situation is influenced by your experiences and might be very different from the perceptions of others.
- Always give others the benefit of the doubt and ask for clarification if you think there is a problem.
- Avoid allowing feelings or emotions to get in the way of how you interpret a situation.

## Fairness

As part of the *PMI Code of Ethics and Professional Conduct*, you have a responsibility to report to the stakeholders, customers, or others any actions or circumstances that could be construed as a conflict of interest. Avoiding conflicts of interest is a part of exhibiting fairness. Fairness includes the following:

- Avoiding favoritism and discrimination
- Avoiding and reporting conflict-of-interest situations
- Maintaining impartiality in your decision-making process

A conflict of interest involves putting your personal interests above the interests of the project or using your influence to cause others to make decisions in your favor without regard for the project outcome. As Figure 1.11 illustrates, any of the following could be construed as conflicts of interest and compromise fairness:

- Not revealing your active associations and affiliations
- Accepting vendor gifts
- Using stakeholder influence for personal reasons
- Participating in discrimination

**FIGURE 1.11** Elements to consider in exhibiting fairness

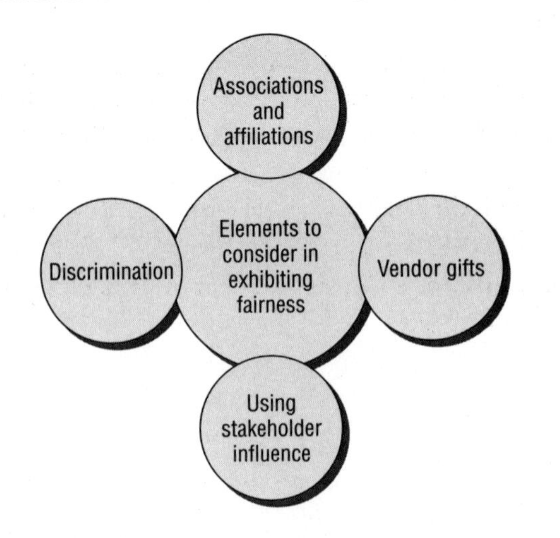

### Associations and Affiliations

In maintaining associations and affiliations, it's important to consider whether any conflicts of interest exist as a result. For instance, sitting on the decision-making committee of a company that wins a bid for a contract relating to your project constitutes a conflict of interest. Even if you do not influence the decision in any way, others will assume that a conflict of interest occurred. This is also the case in personal associations—for instance, awarding a project to a relative, spouse, and so forth.

When these types of situations occur, do the following:

- Inform the project sponsor and the decision committee of the association or affiliation.
- Refrain from participating in the award decision committee.
- In cases of a contract with a personal association, appoint someone else in your organization to administer the contract and make the payments for the work performed.
- Document all of the decisions you make regarding the activities performed by the association and keep them with the project files.

### Vendor Gifts

Vendor gifts are not uncommon. Vendors and suppliers often provide their customers and potential customers with lunches, gifts, ballgame tickets, and the like. However, some professionals are not allowed to accept gifts in excess of a certain dollar amount. This might be driven by company policy or the department manager's policy. Accepting vendor gifts may be construed as a conflict of interest, no matter how small the gift is. It is your responsibility to know the policy on accepting gifts and inform the vendor if they go beyond the acceptable limits. Likewise, do not give inappropriate gifts to others or attempt to influence a decision for personal gain.

Don't accept gifts that might be construed as a conflict of interest. If your organization does not have a policy regarding vendor gifts, set limits for yourself.

### Stakeholder Influence

Some stakeholders have a large amount of authority. It is important not to put your own personal interests above the interests of the project or organization when dealing with powerful stakeholders. This is another potential area for conflict of interest. Keep in mind these guidelines:

- Weigh your decisions with the objectives of the project and the organization in mind—not your own personal gain.
- Do not use stakeholder influence for personal gain, such as receiving a promotion, bonus, or similar reward.

### Discrimination

Maintaining a strong level of fairness is important on many levels, as you have seen so far. In practice, we sometimes fall prey to favoritism without realizing it. When issuing assignments, promotions, and the like, always use strong professional reasoning when making decisions. Choosing someone simply because you enjoy working with them is poor judgment and represents a conflict of interest.

It is also important to be conscious of discriminating against others. Whether based on age, disability, religion, sexual orientation, or personality, discrimination on any level is not acceptable. Remember that the *PMI Code of Ethics and Professional Conduct* also requires you to report this type of behavior exhibited by someone else.

## Honesty

There are several aspects of honesty that you should be aware of and need to uphold (as shown in Figure 1.12):

- Reporting the truth regarding project status and circumstances
- Avoiding deception
- Making only truthful statements
- Being honest about your own experience

**FIGURE 1.12** Elements of honesty

### Reporting Truthfully

Honesty involves information regarding your own background and experience as well as information about the project circumstances. Benefit to the project, and not your personal gain, should be the influencing factor when billing a customer or working on a project. Experiencing a personal gain as a result of your work is positive, but making a project decision simply to result in a personal gain is not.

As a project manager, you are responsible for truthfully reporting all information in your possession to stakeholders, customers, the project sponsor, and the public when required. This includes situations where personal gain is at stake. Always being truthful about the situation and also being up front regarding the project's progress is important.

### Avoiding Deception

At times, being honest can feel difficult. In some cases, it may result in the end of a project, losing a job, missing out on a bonus, or other consequences that impact you and your team personally. Deceiving others and making false statements to prevent unfavorable consequences are not acceptable ways of handling situations. In terms of a project, being up front and telling the truth are necessities.

### Making Truthful Statements

Part of managing a good team involves leading by example. This includes making truthful statements, not just on your PMP application, but in the day-to-day management of a project. In some cases, delicate situations may present themselves, which is where your general management skills come into play. Being truthful doesn't mean being disrespectful, and a project manager must understand how to balance both.

### Representing Personal Experience

Being honest about your personal experience is just as important as not deceiving others. You may have heard that enhancing a resume is a standard practice in business. While creating a resume that greatly highlights your experiences and credentials is recommended, exaggerating experience or clearly embellishing credentials are unacceptable. This is equivalent to making false statements.

Correctly representing yourself naturally applies to your PMP application. Accurately stating your work experience, credentials, and background is an important part of maintaining high integrity.

## Advancing the Industry

Promoting good project management and advancing the industry are part of your role as a PMP. Professional knowledge involves the knowledge of project management practices as well as specific industry or technical knowledge required to complete an assignment. As part of maintaining your credential, you have the following responsibilities:

- To apply project management knowledge to all your projects
- To educate others regarding project management practices by keeping them up-to-date on best practices, training your team members to use the correct techniques, informing stakeholders of correct processes, and then sticking to those processes throughout the course of the project
- To promote interaction among stakeholders

Becoming knowledgeable in project management best practices and staying abreast of the industry are important. Figure 1.13 touches on these responsibilities.

**FIGURE 1.13** Elements of industry advancement

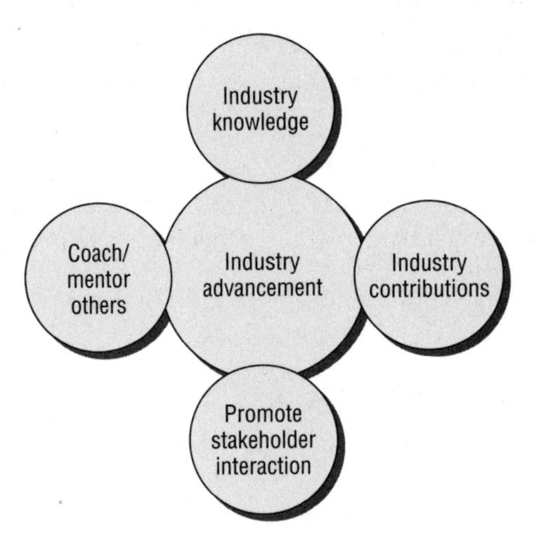

While the PMP exam is not expected to test your knowledge of this topic, the continuing credential requirements will become important toward maintaining your credential.

## Enhancing Your Industry Knowledge

Furthering knowledge in the field of project management (your own and others) is a part of your responsibility as a PMP. Project management is a growing field, which makes staying abreast of project management practices, theories, and techniques an ongoing process. For this reason, PMI requires that, once you achieve your PMP, you maintain the credential by earning a given number of professional development units (PDUs) every three years. PDUs ensure that your knowledge of the industry and project management practices remains current. There are several ways of earning PDUs, such as taking project management classes, reading industry books and magazines, speaking on the subject of project management, and so forth. You can find additional details on PMI's website (`www.PMI.org`).

Here are some ways that you can further and enhance your project management education:

- Join a local PMI chapter.
- Join a PMI Community of Practice (CoP).
- Read PMI's monthly magazine.
- Take educational courses offered through PMI chapters or other project management training companies.

Remember that as a PMP, you should be proactive in looking for ways to further enhance your skills and knowledge of the industry.

## Contributing to the Industry

As a PMP you have the responsibility of not only furthering your own level of knowledge and skill set, but also contributing to the knowledge base itself. This includes actions such as the following:

- Coaching and mentoring others within the field
- Contributing to the knowledge base itself, such as through volunteer efforts, writing articles relating to project management, and conducting research and studies that enhance or add to the industry
- Supporting other project managers

## Promoting Stakeholder Interaction

Within a project, there can be a small handful to dozens of stakeholders. The greater the number of stakeholders, the more complex project communication and control become. Throughout this book, we review how stakeholders tend to have varying interests. Although everyone wants the project to succeed, the agendas may vary. There are several ways to promote interaction among stakeholders:

- Working toward balancing their interests, requirements, and expectations
- Reducing the amount of conflict within the project
- Encouraging collaboration among the departments

Be sure to apply all of the stakeholder management techniques that we will cover later in this book, from within the Project Communications Management Knowledge Area.

# Review Questions

1. The following are characteristics of a project EXCEPT:
   A. It is temporary in nature.
   B. It is continuous.
   C. It is unique.
   D. It has a definitive end date.

2. When is a project considered successful?
   A. All deliverables have been completed.
   B. The phase completion has been approved.
   C. Expectations of all stakeholders have been met.
   D. The customer has provided final payment.

3. You are currently managing multiple projects relating to a high-profile build-out of a major retail center. The project is extensive, with a two-year timeline. Bob is the project manager responsible for the east block of the build-out, while Sally is the project manager responsible for the north block of the build-out. Both Bob and Sally report to you. Your role can best be described as:
   A. Lead project manager
   B. Department manager
   C. Program manager
   D. Portfolio manager

4. Which of the following organizational types is the oldest style of organization?
   A. Functional organization
   B. Projectized organization
   C. Matrix organization
   D. Balanced matrix organization

5. During a planning meeting, a discussion takes place to decide which resource will be assigned to perform an activity that will be fast-tracked. As the project manager, you decide to assign Fred to the activity. What type of organizational structure is this?
   A. Functional organization
   B. Projectized organization
   C. Matrix organization
   D. Balanced matrix organization

6. All of the following are process groups EXCEPT:
   A. Integration
   B. Planning
   C. Executing
   D. Closing

7. During which process group are costs the highest?
   A. Planning
   B. Executing
   C. Monitoring and Controlling
   D. Closing

8. The concept that each element within a cycle is results-oriented and becomes an input into the next cycle describes:
   A. Project management processes
   B. Phases
   C. Project life cycle
   D. Plan-Do-Check-Act cycle

9. All of the following statements are true EXCEPT:
   A. It is the job of the project manager to ensure that information is explicit, clear, and complete.
   B. A project manager must possess a strong knowledge of finance and accounting principles.
   C. Organizational and planning skills are an important skill set for a project manager to possess.
   D. Power is a technique that the project manager may use to influence people.

10. Which of the following BEST describes the level of authority that a project manager has within a functional organization?
    A. High to ultimate authority
    B. Authority to assign resources
    C. Minimal to no authority
    D. Authority to recommend resources

# Answers to Review Questions

1. B. Ongoing operations are continuous without a specific end date, while projects have definite start and end dates.

2. C. Although all the options appear attractive, a project is not considered successful until it has met the expectations of the project stakeholders.

3. C. Programs include groups of related projects managed in a coordinated fashion. Each project under a program is managed by its own project manager, who reports to a project manager responsible either for that area of the program or for the entire program.

4. A. The functional organization type is considered to be the traditional approach to organizing businesses. Functional organizations are centered around specialties, grouped by function, and displayed as a hierarchy.

5. B. Within a projectized organization, project managers have high to ultimate authority, giving them the capability of assigning resources.

6. A. Integration refers to a knowledge area within the *PMBOK® Guide*, which will be addressed in a future chapter. Because the word closely resembles Initiating, the first process group within the project life cycle, they can easily be confused.

7. B. Costs are highest during the Executing process group, since this is where the most time and resources are utilized within a project.

8. D. This describes the Plan-Do-Check-Act cycle, which was originally defined by Walter Shewhart and later modified by Edward Deming.

9. B. While having a strong knowledge of finance and accounting principles would be beneficial to a project manager, only a basic level of knowledge is required.

10. C. Within a functional organization, the functional manager has a higher level of authority over the project manager, who has minimal to none.

Chapter 2

# Initiating the Project

## THE PMP EXAM CONTENT FROM THE INITIATING THE PROJECT PERFORMANCE DOMAIN COVERED IN THIS CHAPTER INCLUDES THE FOLLOWING:

- ✓ **Perform project assessment based upon available information and meetings with the sponsor, customer, and other subject matter experts, in order to evaluate the feasibility of new products or services within the given assumptions and/or constraints.**
- ✓ **Develop the project charter by further gathering and analyzing stakeholder requirements, in order to document project scope, milestones, and deliverables.**
- ✓ **Define the high-level scope of the project based on business and compliance requirements, in order to meet the customer's project expectations.**
- ✓ **Identify and document high-level risks, assumptions, and constraints based on current environment, historical data, and/or expert judgment, in order to identify project limitations and propose an implementation approach.**
- ✓ **Obtain approval for the project charter from the sponsor and customer (if required), in order to formalize the authority assigned to the project manager and gain commitment and acceptance for the project.**
- ✓ **Perform key stakeholder analysis using brainstorming, interviewing, and other data-gathering techniques, in order to ensure expectation alignment and gain support for the project.**

Initiating is the first of the five project management process groups. Initiating acknowledges that a new project (or the next phase in an active project) should begin. This process group culminates in the publication of a project charter and a stakeholder register.

Before diving into the Initiating process group, we'll provide a high-level review of the nine Project Management Knowledge Areas, which is another way of classifying the project management processes. This review offers a high-level look at the Knowledge Areas and insight as to how they are tied into the five process groups.

The Initiating process group accounts for 13 percent of the questions on the PMP exam.

The descriptions, process names, and project management process groups in the nine Knowledge Areas described in the following sections are based on content from *A Guide to the Project Management Body of Knowledge, 4th Edition* (*PMBOK® Guide*).

# Understanding the Project Management Knowledge Areas

Throughout the project life cycle, a project utilizes a collection of processes that can be classified by the five process groups mentioned in Chapter 1, "Project Foundation." They are Initiating, Planning, Executing, Monitoring and Controlling, and Closing. Processes can also be classified into nine categories called the Project Management Knowledge Areas, which bring together processes that have characteristics in common. According to the *PMBOK® Guide*, the nine Knowledge Areas are as follows:

- Project Integration Management
- Project Scope Management
- Project Time Management
- Project Cost Management
- Project Quality Management
- Project Human Resource Management
- Project Communications Management
- Project Risk Management
- Project Procurement Management

Processes within the Knowledge Areas are grouped by commonalities, whereas processes within the project management process groups are grouped in more or less the order in which you perform them. Once again, these are two ways of classifying the same set of project management processes. Processes tend to interact with other processes outside of their own Knowledge Area.

For more detailed information on the Project Management Knowledge Areas, see Chapter 2, "Creating the Project Charter," of *PMP: Project Management Professional Exam Study Guide, 6th Edition* (Sybex, 2011).

## Project Integration Management

The Project Integration Management Knowledge Area consists of six processes. Table 2.1 lists the processes and the process group for each. The Project Integration Management Knowledge Area involves identifying and defining the work of the project and combining, unifying, and integrating the appropriate processes. This is the only Knowledge Area that contains processes across all five of the project management process groups. The processes within Project Integration are tightly linked because they occur continually throughout the project.

**TABLE 2.1** Project Integration Management Knowledge Area

| Process Name | Project Management Process Group |
|---|---|
| Develop Project Charter | Initiating |
| Develop Project Management Plan | Planning |
| Direct and Manage Project Execution | Executing |
| Monitor and Control Project Work | Monitoring and Controlling |
| Perform Integrated Change Control | Monitoring and Controlling |
| Close Project or Phase | Closing |

The Project Integration Management Knowledge Area involves the following:

- Identifying and defining the work of the project
- Combining, unifying, and integrating the appropriate processes
- Managing customer and stakeholder expectations and meeting stakeholder requirements

The following tasks are typical of those performed during work covered by the Project Integration Management Knowledge Area:

- Making choices about where to concentrate resources
- Making choices about where to expend daily efforts
- Anticipating potential issues
- Dealing with issues before they become critical
- Coordinating the work for the overall good of the project
- Making trade-offs among competing objectives and alternatives

Figure 2.1 shows the highlights of the Knowledge Area's purpose.

**FIGURE 2.1** Project Integration Management

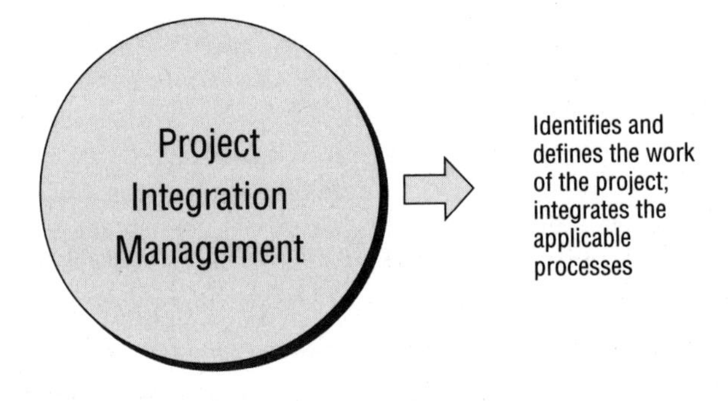


## Project Scope Management

The Project Scope Management Knowledge Area consists of five processes. Table 2.2 lists the processes and the process group for each. These processes are interactive and define and control what is strategically included versus intentionally excluded from the project scope.

**TABLE 2.2** Project Scope Management

| Process Name | Project Management Process Group |
|---|---|
| Collect Requirements | Planning |
| Define Scope | Planning |
| Create WBS | Planning |
| Verify Scope | Monitoring and Controlling |
| Control Scope | Monitoring and Controlling |

The Project Scope Management Knowledge Area encompasses the following:

**Project Scope** Project scope, according to the *PMBOK® Guide*, identifies the work that needs to be accomplished to deliver a product, service, or result with the specified features and functionality. This is typically measured against the project management plan.

**Product Scope** Product scope addresses the features and functionality that characterize the product, service, or result of the project. This is measured against the product specifications to determine successful completion or fulfillment. Product scope entails the following:

- Detailing the requirements of the product of the project
- Verifying those details using measurement techniques
- Creating a project scope management plan
- Creating a work breakdown structure (WBS)
- Controlling changes to these processes

The following tasks are typical of those performed during work covered by the Project Scope Management Knowledge Area:

- Collecting the project requirements
- Defining the overall project scope
- Measuring the project against the project plan, project scope statement, and WBS
- Measuring the completion of a product against the product requirements

Keep in mind that the Project Scope Management Knowledge Area is concerned with making sure the project includes *all* the work required to complete the project—and *only* the work required to complete the project successfully.

Figure 2.2 shows the highlights of the Knowledge Area's purpose.

**FIGURE 2.2** Project Scope Management

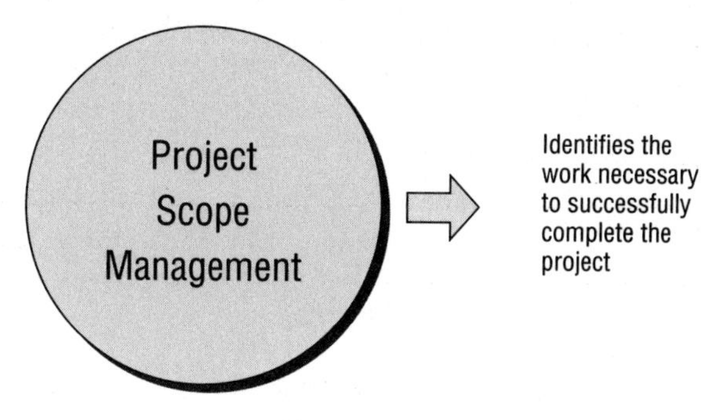

## Project Time Management

The Project Time Management Knowledge Area consists of six processes. Table 2.3 lists the processes and the process group for each. On small projects, the Sequence Activities, Estimate Activity Durations, and Develop Schedule processes are completed as one activity.

**TABLE 2.3** Project Time Management

| Process Name | Project Management Process Group |
|---|---|
| Define Activities | Planning |
| Sequence Activities | Planning |
| Estimate Activity Resources | Planning |
| Estimate Activity Durations | Planning |
| Develop Schedule | Planning |
| Control Schedule | Monitoring and Controlling |

The Project Time Management Knowledge Area involves the following:

- Keeping project activities on track and monitoring those activities against the project plan
- Ensuring that the project is completed on time

The flow of the Project Time Management Knowledge Area is very intuitive. First, the project activities are defined and sequenced. Next, the resources are estimated for each activity, along with the estimated duration. Finally, the project schedule is developed and controlled as the project work moves forward.

Figure 2.3 shows the highlights of the Knowledge Area's purpose.

**FIGURE 2.3** Project Time Management

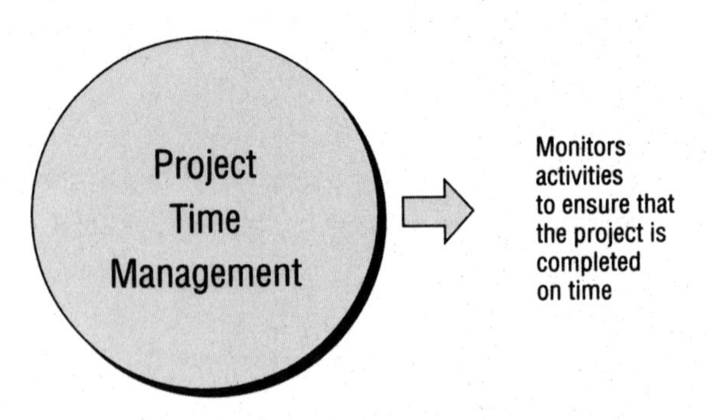

## Project Cost Management

The Project Cost Management Knowledge Area consists of three processes. Table 2.4 lists the processes and the process group for each. For small projects, the Estimate Costs process and Determine Budget process may be combined into a single process. Remember that your ability to influence cost is greatest at the beginning of the project. Minimal resources have been used, in comparison with other stages. Therefore, changes made during these processes have a greater impact on cost reduction, and your opportunities to effect change decline as you move deeper into the project.

**TABLE 2.4** Project Cost Management

| Process Name | Project Management Process Group |
|---|---|
| Estimate Costs | Planning |
| Determine Budget | Planning |
| Control Costs | Monitoring and Controlling |

The Project Cost Management Knowledge Area involves the following:

- Establishing cost estimates for resources
- Establishing budgets
- Ensuring that the project is executed within the approved budget
- Improving the quality of deliverables through two techniques: life-cycle costing and value engineering

Life-cycle costing and value engineering are used to improve deliverables. The following is a brief overview of these two techniques:

**Life-Cycle Costing** Life-cycle costing is an economic evaluation technique that determines the total cost of not only the project (temporary costs) but also owning, maintaining, and operating something from an operational standpoint after the project is completed and turned over.

**Value Engineering** Value engineering is a technique that looks for alternative product ideas to make certain the team is applying the lowest cost for the same or better-quality results. So, for example, if the original engineer suggests a custom part for a step of the production process, a value engineering review might suggest an alternative that can use a readily available, lower-cost standard part that will produce the same output for a better value.

These are typical tasks performed during work covered by the Project Cost Management Knowledge Area:

- Developing the cost management plan
- Considering the effects of project decisions on cost
- Considering the information requirements of stakeholders

Figure 2.4 shows the highlights of the Knowledge Area's purpose.

**FIGURE 2.4** Project Cost Management

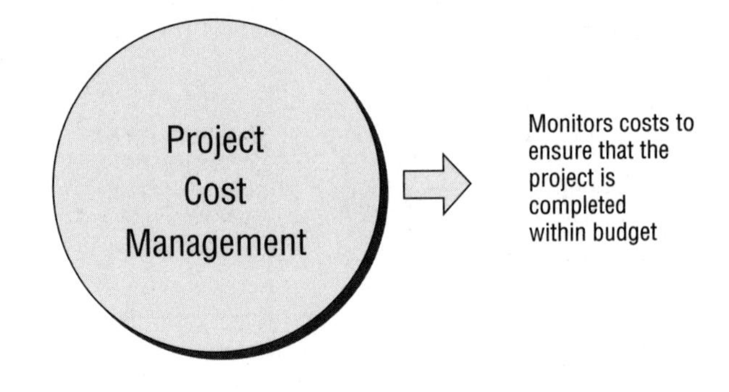


## Project Quality Management

The Project Quality Management Knowledge Area consists of three processes. Table 2.5 lists the processes and the process group for each.

**TABLE 2.5** Project Quality Management

| Process Name | Project Management Process Group |
|---|---|
| Plan Quality | Planning |
| Perform Quality Assurance | Executing |
| Perform Quality Control | Monitoring and Controlling |

The Project Quality Management Knowledge Area involves the following:

- Ensuring that the project meets the requirements that it was undertaken to produce
- Converting stakeholder needs, wants, and expectations into requirements
- Focusing on product quality and the quality of the project management processes used
- Measuring overall performance
- Monitoring project results and comparing them with the quality standards established in the project planning process

The following tasks are typical of those performed during work covered by the Project Quality Management Knowledge Area:

- Implementing the quality management system through the policy, procedures, and quality processes
- Improving the processes
- Verifying that the project is compatible with ISO standards

Figure 2.5 shows the highlights of the Knowledge Area's purpose.

**FIGURE 2.5** Project Quality Management

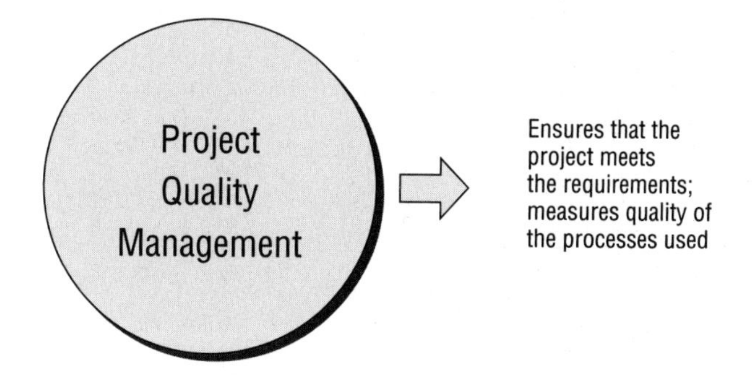


## Project Human Resource Management

The Project Human Resource Management Knowledge Area consists of four processes. These processes organize, develop, and manage the project team. Table 2.6 lists the processes and the process group for each. Because the makeup of each team is different and the stakeholders involved in the various stages of the project might change, project managers use different techniques at different times throughout the project to manage these processes.

**TABLE 2.6** Project Human Resource Management

| Process Name | Project Management Process Group |
|---|---|
| Develop Human Resource Plan | Planning |
| Acquire Project Team | Executing |
| Develop Project Team | Executing |
| Manage Project Team | Executing |

The Project Human Resource Management Knowledge Area involves the following:

- All aspects of people management and personnel interaction
- Ensuring that the human resources assigned to the project are used in the most effective way possible
- Practicing good project management by knowing when to enact certain skills and communication styles based on the situation

The Project Human Resource Knowledge Area also includes enhancing the skill and efficiency of the project team, which, in turn, improves project performance.

Keep in mind that the project management team and project team are different. The project management team is the group of individuals responsible for planning, controlling, and closing activities. They are considered to be the leadership team of the project. The project team is made up of all the individuals that have assigned roles and responsibilities for completing the project. They are also referred to as the project staff.

Figure 2.6 shows the highlights of the Knowledge Area's purpose.

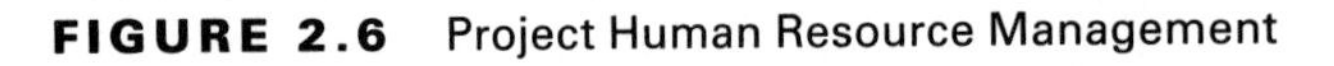

**FIGURE 2.6** Project Human Resource Management

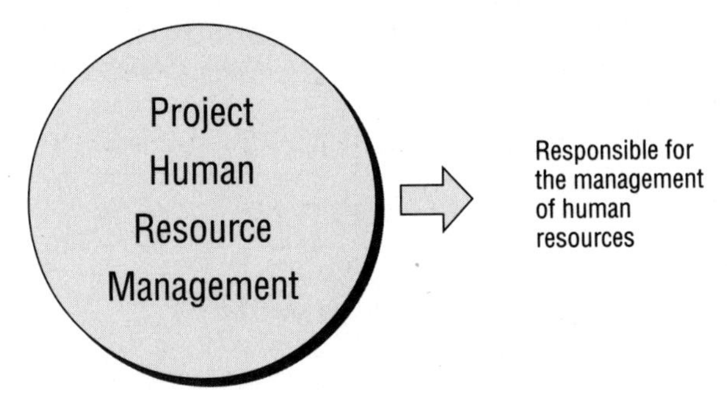


## Project Communications Management

The Project Communications Management Knowledge Area consists of five processes. Together, these processes connect people and information for successful communication throughout the project. Table 2.7 lists the processes and the process group for each.

**TABLE 2.7** Project Communications Management

| Process Name | Project Management Process Group |
|---|---|
| Identify Stakeholders | Initiating |
| Plan Communications | Planning |
| Distribute Information | Executing |
| Manage Stakeholder Expectations | Executing |
| Report Performance | Monitoring and Controlling |

The Project Communications Management Knowledge Area involves the following:

- Ensuring that all project information is collected, documented, archived, and disposed of at the proper time.
- Distributing and sharing information with stakeholders, management, and project team members.
- Archiving information after project closure to be used as reference for future projects; this information is known as historical information.

As mentioned in Chapter 1 of this book, a good project manager spends up to 90 percent of their time communicating. Overall, much of the project manager's work involves managing project communication.

Figure 2.7 shows the highlights of the Knowledge Area's purpose.

**FIGURE 2.7** Project Communications Management

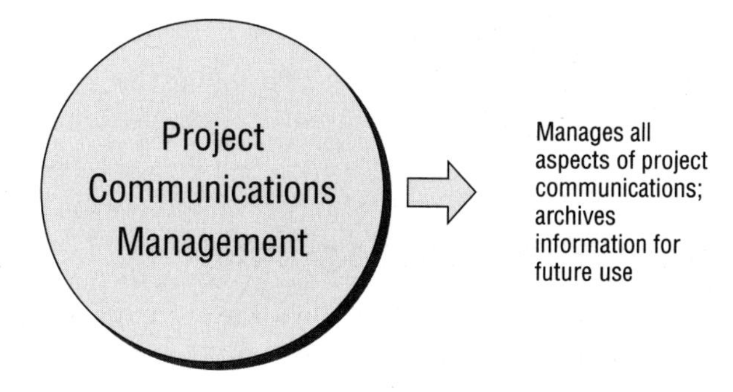


## Project Risk Management

The Project Risk Management Knowledge Area consists of six processes, some of which may be combined into one step. Table 2.8 lists the processes and the process group for each. Risks include both threats to and opportunities within the project.

**TABLE 2.8** Project Risk Management

| Process Name | Project Management Process Group |
|---|---|
| Plan Risk Management | Planning |
| Identify Risks | Planning |
| Perform Qualitative Risk Analysis | Planning |

**TABLE 2.8** Project Risk Management *(continued)*

| Process Name | Project Management Process Group |
|---|---|
| Perform Quantitative Risk Analysis | Planning |
| Plan Risk Responses | Planning |
| Monitor and Control Risk | Monitoring and Controlling |

The Project Risk Management Knowledge Area involves the following:

- Identifying, analyzing, and planning for potential risks, both positive and negative, that may impact the project
- Minimizing the probability and impact of negative risks, and increasing the probability and impact of positive risks
- Identifying the positive consequences of risk and exploiting them to improve project objectives or discover efficiencies that may improve project performance

According to the *PMBOK® Guide*, a risk is an uncertain event or condition that has a positive or negative effect on a project objective. Every project has some level of uncertainty and, therefore, some level of risk. Keep in mind that a risk is different from an issue. A risk may or may not occur. But when a negative risk materializes and impacts the project, it becomes an issue that must be handled.

Figure 2.8 shows the highlights of the Knowledge Area's purpose.

**FIGURE 2.8** Project Risk Management

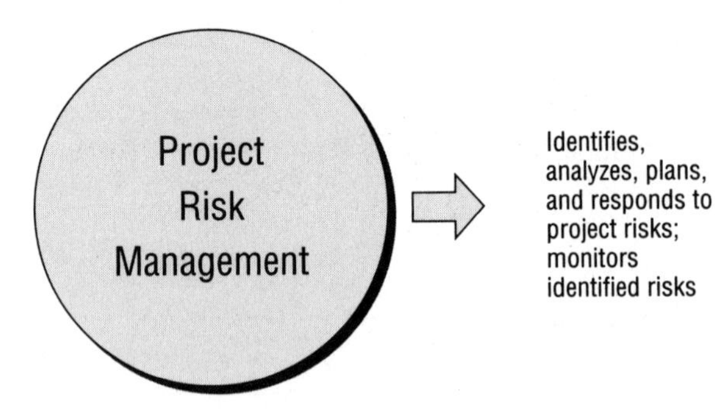


## Project Procurement Management

The Project Procurement Management Knowledge Area consists of four processes. Table 2.9 lists the processes and the process group for each. While reviewing the procurement processes, remember that you are viewing the process from the perspective of the buyer, and that sellers are external to the project team. Together, these four processes allow you to manage the purchasing activities of the project and the life cycle of the procurement contracts.

**TABLE 2.9** Project Procurement Management

| Process Name | Project Management Process Group |
|---|---|
| Plan Procurements | Planning |
| Conduct Procurements | Executing |
| Administer Procurements | Monitoring and Controlling |
| Close Procurements | Closing |

The Project Procurement Management Knowledge Area involves purchasing goods or services from vendors, contractors, suppliers, and others outside the project team.

Keep in mind that a contract can be simple or complex; contracts can be tailored to the needs of the project. The *PMBOK® Guide* defines a contract as an agreement that binds the seller to provide the specified products or services; it also obligates the buyer to compensate the seller as specified.

Figure 2.9 shows the highlights of the Knowledge Area's purpose.

**FIGURE 2.9** Project Procurement Management

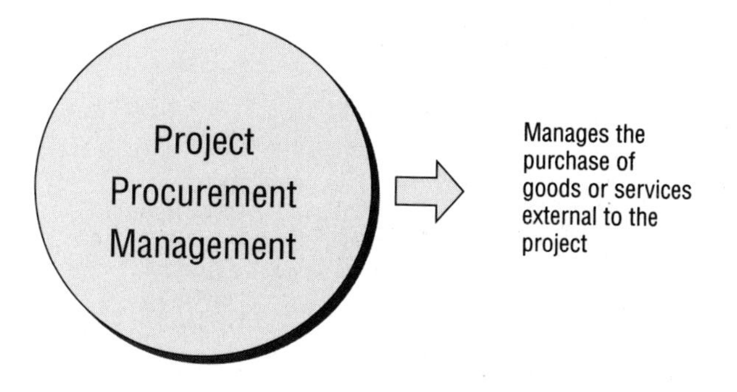


## Exam Essentials

**Be able to name the nine Project Management Knowledge Areas.** The nine Project Management Knowledge Areas are Project Integration Management, Project Scope Management, Project Time Management, Project Cost Management, Project Quality Management, Project Human Resource Management, Project Communications Management, Project Risk Management, and Project Procurement Management.

# Performing a Project Assessment

Before an organization decides to embark on a project, several considerations must take place, such as an assessment of the various potential projects. As part of performing a project assessment, management will need to evaluate existing needs and demands, perform feasibility studies, and use project selection methods to help in making good decisions.

Project assessment should entail a careful evaluation of information carried out through meetings with the sponsor, customer, and other subject matter experts. This group of key individuals will then determine whether the new products or services are feasible under the existing assumptions and constraints.

## Initiating a Project

Project initiation is the formal recognition that a new project, or the next phase in an existing project, should begin and resources may be committed to the project. A project may come about as a result of a need, demand, opportunity, or problem. Once those needs and demands are identified, the next logical step may include performing a feasibility study to determine the viability of the project.

For more detailed information on initiating a project, including a closer look at the project selection methods and a case study, see Chapter 2 of *PMP: Project Management Professional Exam Study Guide, 6th Edition.*

### Identifying Needs and Demands

Needs and demands represent opportunities, business requirements, or problems that need to be solved. Management must decide how to respond to these needs and demands, which will, more often than not, initiate new projects. According to the *PMBOK® Guide*, projects come about as a result of one of the following seven needs or demands:

**Market Demand** The demands of the marketplace can drive the need for a project. Some companies must initiate projects in order to take advantage of temporary and long-term market changes.

**Organizational Need** An organization may respond to an internal need that could eventually affect the bottom line. For example, this may include addressing company growth or even the need to downsize.

**Customer Request** A new project may emerge as a result of internal or external customer requests.

**Technological Advance** New technology often requires companies to revamp their products as a way of taking advantage of the latest technology.

**Legal Requirement** Both private industry and government agencies generate new projects as a result of laws passed during every legislative season.

**Ecological Impact** Many organizations today are undergoing a greening effort to reduce energy consumption, save fuel, reduce their carbon footprint, and so on.

**Social Need** Projects arise out of social needs, such as a developing country that offers medical supplies and vaccination in response to a fast-spreading disease.

### Conducting a Feasibility Study

Some organizations require that a feasibility study take place prior to making a final decision about starting a project. Feasibility studies may be conducted as separate projects, as subprojects, or as the first phase of a project.

When evaluating the feasibility of new products or services, it's generally a good idea to meet with the sponsor, customer, and other subject matter experts. Here are some of the reasons that a feasibility study may be undertaken:

- To determine whether the project is viable
- To determine the probability of success
- To examine the viability of the product, service, or result of the project
- To evaluate technical issues related to the project
- To determine whether the technology proposed is feasible, reliable, and easily assimilated into the existing technology structure

## Selecting a Project

As shown in Figure 2.10, there are a variety of selection methods an organization may choose to utilize. Selection methods help organizations decide among alternative projects and determine the tangible benefits to the company of choosing or not choosing a project. Project selection methods are also used to evaluate and choose between alternative ways to implement the project.

**FIGURE 2.10** Project selection methods

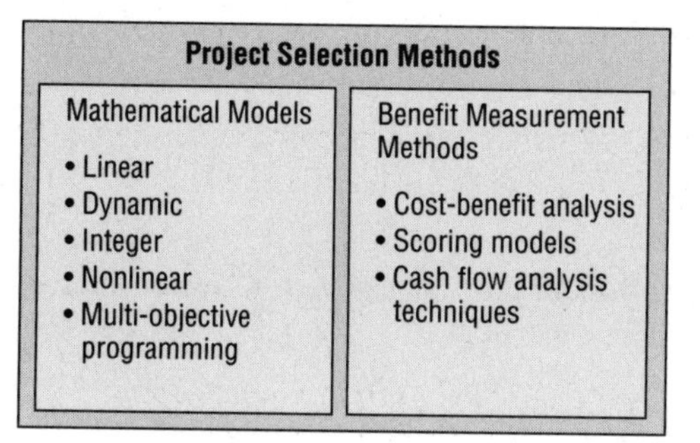

Although project selection is considered to be out of the scope of a project manager's role, the project selection methods we'll talk about next are ones you should be familiar with for the exam.

Depending on the organization, a steering committee may be responsible for project review, selection, and prioritization. A *steering committee* is a group comprising senior managers and, sometimes, mid-level managers who represent each of the functional areas in the organization.

There are generally two categories of selection methods:

- Mathematical models (also known as constrained optimization methods)
- Benefit measurement methods (also known as decision models)

## Mathematical Models

For the exam, simply know that mathematical models, also known as constrained optimization methods, use the following in the form of algorithms:

- Linear
- Dynamic
- Integer
- Nonlinear
- Multi-objective programming

## Benefit Measurement Methods

Benefit measurement methods, also known as decision models, employ various forms of analysis and comparative approaches to make project decisions. These methods include comparative approaches, such as these:

- Cost-benefit analysis
- Scoring models
- Cash flow analysis techniques

When applying project selection methods, you can use a benefit measurement method alone or in combination with others to come up with a selection decision.

Let's take a closer look at the benefit measurement methods.

### Cost-Benefit Analysis

Cost-benefit analysis, also known as cost analysis or benefit analysis, compares the cost to produce the product, service, or result of the project with the benefit that the organization will receive as a result of executing the project.

### Scoring Models

For weighted scoring models, the project selection committee decides on the criteria that will be used on the scoring model. Each of these criteria is then assigned a weight depending on its importance to the project committee. More-important criteria should carry a higher weight than less-important criteria. Each project is then rated on a numerical scale, with the higher number being the more desirable outcome to the company and the lower number having the opposite effect. This rating is then multiplied by the weight of the criteria factor and added to other weighted criteria scores for a total weighted score. The project with the highest overall weighted score is the best choice.

For example, if Project A contains a weighted score of 31, Project B contains a weighted score of 42, and Project C contains a weighted score of 40, which project would you choose? The correct answer is Project B, since it has the highest weighted score.

### Cash Flow Analysis Techniques

The final benefit measurement methods involve a variety of cash flow analysis techniques:

- Payback period
- Discounted cash flows
- Net present value
- Internal rate of return

Figure 2.11 provides an overview of the cash flow analysis techniques.

**FIGURE 2.11** Overview of cash flow analysis techniques

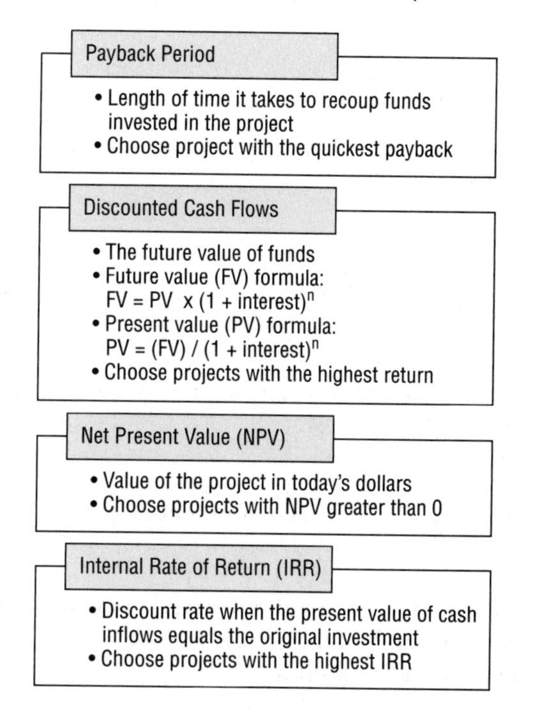

Let's take a closer look at these techniques.

### PAYBACK PERIOD

The *payback period* is the length of time it takes the company to recoup the initial costs of producing the product, service, or result of the project. This method compares the initial investment to the cash inflows expected over the life of the product, service, or result.

#### Payback Period Example

The initial investment on a project is $200,000, with expected cash inflows of $25,000 per quarter every quarter for the first two years and $50,000 per quarter from then on. The payback period is two years and can be calculated as follows:

Initial investment = $200,000

Cash inflows = $25,000

Time frame = 4 quarters

Total cash inflows per year = $100,000 (4 quarters x cash inflows of $25,000)

Initial investment ($200,000) – year 1 inflows ($100,000) = $100,000 remaining balance

Year 1 inflows remaining balance – year 2 inflows = $0

Total cash flow year 1 and year 2 = $200,000

The payback is reached in two years.

The fact that inflows are $50,000 per quarter starting in year 3 makes no difference because payback is reached in two years.

The payback period is the least precise of all the cash flow calculations because it does not consider the time value of money.

### DISCOUNTED CASH FLOWS

Money received in the future is worth less than money received today. Therefore, you can calculate what the value of funds will be in the future by using the following formula:

$$FV = PV\,(1 + i)^n$$

This formula says that the future value (FV) of the investment equals the present value (PV) times (1 plus the interest rate) raised to the value of the number of time periods ($n$) the interest is paid.

**Future Value Example**

What is the value of $2,000 three years from today, at 5 percent interest per year?

$FV = \$2{,}000(1.05)^3$

$FV = \$2{,}000(1.157625)$

$FV = \$2{,}315.25$

The discounted cash flow technique compares the value of the future cash flows of the project to today's dollars. It is literally the reverse of the FV formula. To calculate discounted cash flows, you need to know the value of the investment in today's terms, or the present value (PV). PV is calculated using the following formula:

$$PV = FV\ (1 + i)^n$$

**Planned Value Example**

If $2,315.25 is the value of the cash flow three years from now, what is the value today given a 5 percent interest rate?

$PV = \$2{,}315.25 \div (1 + 0.05)^3$

$PV = \$2{,}315.25 \div 1.157625$

$PV = \$2{,}000$

$2,315.25 three years from now is worth $2,000 today.

To calculate discounted cash flow for the projects you are comparing for selection purposes, apply the PV formula, and then compare the discounted cash flows of all the projects against each other to make a selection.

**Calculating Project Value Example**

Project A is expected to make $100,000 in two years, and Project B is expected to make $120,000 in three years. If the cost of capital is 12 percent, which project should you choose?

Using the PV formula shown previously, calculate each project's worth:

PV of Project A = $79,719

PV of Project B = $85,414

Project B is the project that will return the highest investment to the company and should be chosen over Project A.

#### NET PRESENT VALUE

*Net present value (NPV)* allows you to calculate an accurate value for the project in today's dollars. NPV works like discounted cash flows in that you bring the value of future monies received into today's dollars. With NPV, you evaluate the cash inflows using the discounted cash flow technique applied to each period the inflows are expected instead of in one sum. The total present value of the cash flows is then deducted from your initial investment to determine NPV. NPV assumes that cash inflows are reinvested at the cost of capital.

Here are some important notes on NPV calculations:

- If the NPV calculation is greater than zero, accept the project.
- If the NPV calculation is less than zero, reject the project.
- Projects with high returns early in the project are better projects than projects with lower returns early in the project.

#### INTERNAL RATE OF RETURN

*Internal rate of return (IRR)* is the discount rate when the present value of the cash inflows equals the original investment. When choosing between projects or when choosing alternative methods of doing a project, projects with higher IRR values are generally considered better than projects with low IRR values.

Keep the following in mind:

- IRR is the discount rate when NPV equals zero.
- IRR assumes that cash inflows are reinvested at the IRR value.
- You should choose projects with the highest IRR value.

## Exam Essentials

**Be able to distinguish between the seven needs or demands that bring about project creation.** The seven needs or demands that bring about project creation are market demand, organizational need, customer requests, technological advances, legal requirements, ecological impacts, and social needs.

**Be able to define decision models.** Decision models are project selection methods that are used prior to the Develop Project Charter process to determine the viability of the project. Decision models include benefit measurement methods and mathematical models.

**Be familiar with payback period.** Payback period is the amount of time it will take the company to recoup its initial investment in the product of the project. It's calculated by adding up the expected cash inflows and comparing them to the initial investment to determine how many periods it takes for the cash inflows to equal the initial investment.

**Understand the concept of NPV and IRR.** Projects with an NPV greater than zero should be accepted, and those with an NPV less than zero should be rejected. Projects with high IRR values should be accepted over projects with lower IRR values. IRR is the discount rate when NPV is equal to zero and IRR assumes reinvestment at the IRR rate.

# Defining the High-Level Project Scope

During the Initiating process group, the high-level project scope is defined and documented. As the project moves into the Planning process group, the high-level scope is further elaborated and defined. This is typically based on the business and compliance requirements and is documented to meet the customer's project expectations. According to the *PMBOK® Guide*, the high-level project scope is contained within the project charter.

For more information on the project charter, see the section titled "Developing the Project Charter" within this chapter.

**Exam Essentials**

**Be able to identify where the high-level scope is documented and what it is based on.** The high-level scope is defined and documented within the project charter and is based on the business and compliance requirements.

# Identifying High-Level Risks, Assumptions, and Constraints

The Initiating process group focuses on documenting not only the high-level project scope, as mentioned previously, but also high-level risks, assumptions, and constraints. This information will be documented within the project charter and will be largely based on the existing environment, historical information, and expert judgment.

As you may have noticed, this process group is concerned with setting a foundation and common understanding of what the project is setting out to achieve. The Planning process group will later expand on this high-level information by working out the granular details.

For more information on the project charter, see the section titled "Developing the Project Charter" within this chapter.

**Exam Essentials**

**Be able to identify where the high-level risks, assumptions, and constraints are documented and what they are based on.** The high-level risks, assumptions, and constraints are defined and documented within the project charter and are largely based on the existing environment, historical information, and expert judgment.

# Developing the Project Charter

Creating the *project charter* is an important initial step to beginning any project because it formally initiates the project within an organization and gives authorization for resources to be committed to the project. The creation of this document involves gathering and analyzing stakeholder requirements, which will lead the way for properly identifying and documenting the project and product scope, milestones, and deliverables.

The project charter document is created out of the first process within the Project Integration Knowledge Area, called Develop Project Charter. This document attempts to not only formally recognize the project, but also identify project limitations and propose an implementation approach.

The project charter documents the name of the project manager and gives that person the authority to assign organizational resources to the project. It also documents the business need, justification, and impact; describes the customer's requirements; sets stakeholder expectations; and ties the project to the ongoing work of the organization. The project is officially authorized when the project charter is signed.

Figure 2.12 shows the inputs, tools and techniques, and outputs of the Develop Project Charter process.

**FIGURE 2.12** Develop Project Charter process

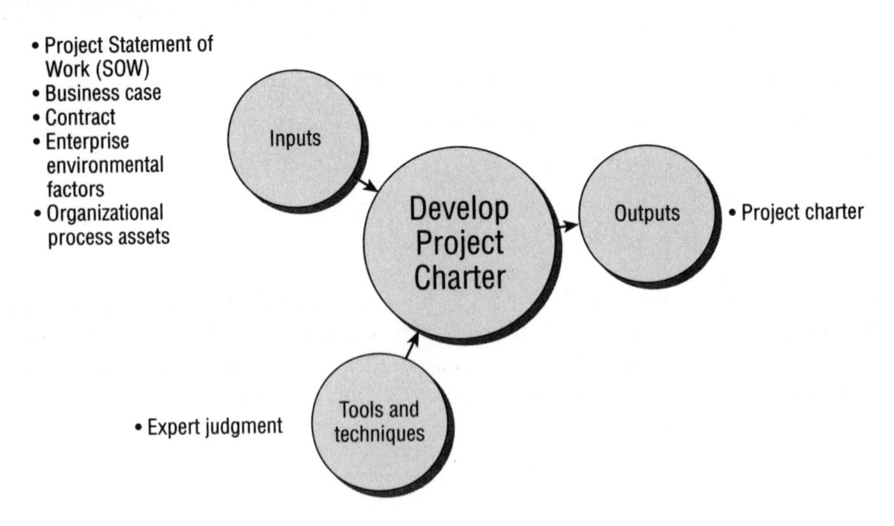

For more detailed information on the project charter, see Chapter 2 of *PMP: Project Managment Professional Exam Study Guide, 6th Edition.*

## Inputs of the Develop Project Charter Process

For the exam, you should know the following inputs of the Develop Project Charter process:

- Project statement of work (SOW)
- Business case
- Contract
- Enterprise environmental factors
- Organizational process assets

Let's look at each of these inputs more closely.

**Project Statement of Work** The project statement of work (SOW) describes the product, service, or result the project was undertaken to complete. When the project is internal, either the project sponsor or the initiator of the project typically writes this document. When the project is external to the organization, it is the buyer that typically writes the SOW. According to the *PMBOK® Guide*, a project SOW should contain or consider all of the following elements:

- Business need
- Product scope description
- Strategic plan

**Business Case** The purpose of a business case is to understand the business need for the project and determine whether the investment in the project is worthwhile. Performing a feasibility study is a great first step in building the business case. Most business case documents contain the following items:

- Description of the business need for the project
- Description of any special requirements that must be met
- Description of alternative solutions
- Description of the expected results of each alternative solution
- Recommended solution

**Contract** The contract input is applicable only when the organization you are working for is performing a project for a customer external to the organization. The contract is used as an input to this process because it typically documents the conditions under which the project will be executed, the time frame, and a description of the work.

**Enterprise Environmental Factors** Enterprise environmental factors refer to the factors outside the project that may have significant influence on the success of the project. According to the *PMBOK® Guide*, the environmental factors, in relation to this process, include but are not limited to the following items:

- Governmental or industry standards, which include elements such as regulatory standards and regulations, quality standards, product standards, and workmanship standards
- Organizational infrastructure, which refers to the organization's facilities and capital equipment
- Marketplace conditions, referring to the supply-and-demand theory along with economic and financial factors

**Organizational Process Assets** Organizational process assets are the organization's policies, guidelines, procedures, plans, approaches, or standards for conducting work, including project work.

These assets include a wide range of elements that might affect several aspects of the project. In relation to this process, organizational process assets refer to the following items:

- Processes, policies, and procedures of the organization
- Corporate knowledge base

## Tools and Techniques of the Develop Project Charter Process

The Develop Project Charter process has only one tool and technique that you should be familiar with: expert judgment. The concept behind expert judgment is to rely on individuals, or groups of individuals, who have training, specialized knowledge, or skills in the areas you're assessing. These individuals may be stakeholders, consultants, other experts in the organization, subject matter experts, the project management office (PMO), industry experts, or technical or professional organizations.

In addition, you should be familiar with who is considered to be a key stakeholder. Key project stakeholders include the following individuals:

**Project Manager** The project manager is the person who assumes responsibility for the success of the project. The project manager should be identified as early as possible in the project and ideally should participate in writing the project charter. The following are examples of what the role of project manager entails:

- Project planning and then executing and managing the work of the project
- Setting the standards and policies for the projects on which they work
- Establishing and communicating the project procedures to the project team and stakeholders
- Identifying activities and tasks, resource requirements, project costs, project requirements, performance measures, and more
- Keeping all stakeholders and other interested parties informed

**Project Sponsor** The project sponsor is usually an executive in the organization who has the power and authority to make decisions and settle disputes or conflicts regarding the project. The sponsor takes the project into the limelight, gets to call the shots regarding project outcomes, and funds the project. The project sponsor should be named in the project charter and identified as the final authority and decision maker for project issues.

**Project Champion** The project champion is usually someone with a great deal of technical expertise or industry knowledge regarding the project who helps focus attention on the project from a technical perspective. Unlike the sponsor, the project champion doesn't necessarily have a lot of authority or executive powers.

**Functional Managers** Project managers must work with and gain the support of functional managers in order to complete the project. Functional managers fulfill the administrative duties of the organization, provide and assign staff members to projects, and conduct performance reviews for their staff.

## Outputs of the Develop Project Charter Process

The Develop Project Charter process results in a single, but critical, output: the project charter.

According to the *PMBOK® Guide*, a useful and well-documented project charter should include at least these elements:

- Purpose or justification for the project
- Project objectives that are measurable
- High-level list of requirements
- High-level description of the project
- High-level list of risks
- Summary milestone schedule
- Summary budget
- Criteria for project approval
- Name of the project manager and their authority levels
- Name of the sponsor (or authorizer of the project) and their authority levels

### Exam Essentials

**Be able to list the Develop Project Charter inputs.** The inputs for Develop Project Charter are project statement of work, business case, contract, enterprise environmental factors, and organizational process assets.

**Be able to describe the purpose of the business case.** The purpose of a business case is to understand the business need for the project and determine whether the investment in the project is worthwhile.

# Obtaining Project Charter Approval

The project charter isn't complete until sign-off has been received from the project sponsor, senior management, and key stakeholders. Sign-off indicates that the document has been read by those signing it and that they agree with the contents and are on board with the project. The signature signifies the formal authorization and acceptance of the project. Acceptance of the charter is also important to formalize the assignment of the project manager and their level of authority.

The last step in this process is publishing the charter. Publishing, in this case, simply means distributing a copy of the project charter to the key stakeholders, the customer, the management team, and others who might be involved with the project. Publication can take several forms, including printed format or electronic format distributed via the company email system or the company intranet.

### Exam Essentials

**Be able to describe the importance of the project charter.** The project charter is the document that officially recognizes and acknowledges that a project exists. The charter authorizes the project to begin. It authorizes the project manager to assign resources to the project and documents the business need and justification. The project charter describes the customer's requirements and ties the project to the ongoing work of the organization.

# Performing Key Shareholder Analysis

Identifying stakeholders and performing analysis on key stakeholders is an important part of setting the stage for a successful project outcome. Stakeholder analysis is carried out through the Identify Stakeholders process and involves the use of brainstorming, interviewing, and several data-gathering techniques.

By performing stakeholder identification and analysis, the project manager can better document stakeholder expectations and ensure that the expectations are aligned with the project objectives. It also serves as a way of gaining project support.

Project stakeholders are those people or organizations who have a vested interest in the outcome of the project. They have something to either gain or lose as a result of the project, and they have the ability to influence project results.

Identify Stakeholders is the first process of the Project Communications Management Knowledge Area and part of the Initiating process group. This process involves identifying and documenting all the stakeholders on the project, including their interests, impact, and potential negative influences on the project. Stakeholder identification should occur as early as possible in the project and continue throughout its life cycle.

Figure 2.13 shows the inputs, tools and techniques, and outputs of the Identify Stakeholders process.

**FIGURE 2.13** Identify Stakeholders process

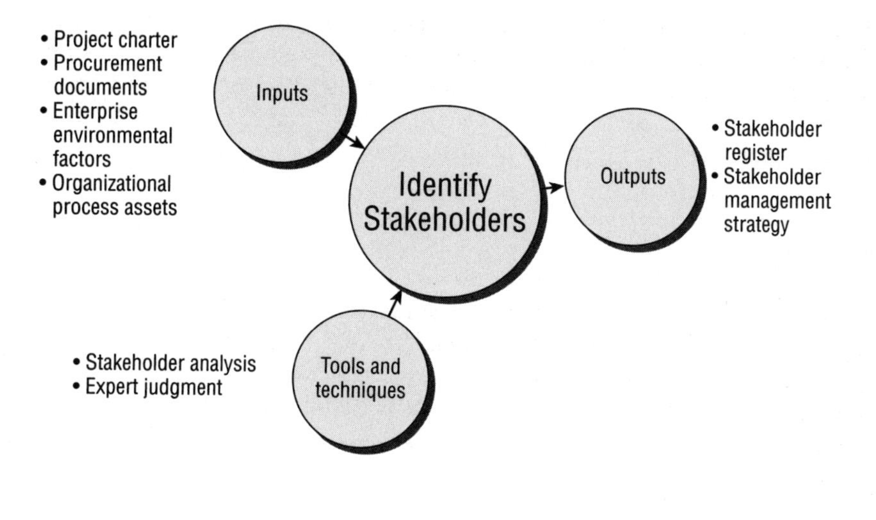

For more detailed information on the Identify Stakeholders process, see Chapter 2 of *PMP: Project Management Professional Exam Study Guide, 6th Edition.*

## Inputs of the Identify Stakeholders Process

There are four inputs within the Identify Stakeholders process:

**Project Charter** Within this process, the project charter provides the list of stakeholders along with other individuals or organizations that are affected by the project.

**Procurement Documents** When the project is procured, the list of key stakeholders and other relevant individuals or organizations may be obtained through the procurement documents.

**Enterprise Environmental Factors** The following enterprise environmental factors are utilized within this process:

- Company culture
- Organizational structure
- Governmental or industry standards

**Organizational Process Assets** The following organizational process assets pertain to this process:

- Stakeholder register templates
- Lessons learned
- Stakeholder register from previous projects

## Tools and Techniques of the Identify Stakeholders Process

The tools and techniques of Identify Stakeholders are stakeholder analysis and expert judgment.

**Stakeholder Analysis** During stakeholder analysis, you'll want to identify the influences stakeholders have in regard to the project and understand their expectations, needs, and desires. From there, you'll derive more specifics regarding the project goals and deliverables. Stakeholders are often concerned with their own interests and what they have to gain or lose from the project.

According to the *PMBOK® Guide*, three steps are involved in stakeholder analysis:

1. Identify all potential stakeholders and capture general information about them, such as the department they work in, contact information, knowledge levels, and influence levels.
2. Identify the potential impact or support each may have to the project, and then classify them according to impact so that you can devise a strategy to deal with them. The *PMBOK® Guide* lists four classification models for classifying the power and influence of each stakeholder on a simple four-square grid:
   - Power/interest grid
   - Power/influence grid
   - Influence/impact grid
   - Salience model (which charts stakeholder power, urgency, and legitimacy)
3. Assess the reaction or responses of the stakeholder to various situations. This will assist the project manager in managing the stakeholders.

**Expert Judgment** Expert judgment is used in combination with stakeholder analysis to ensure that stakeholder identification is comprehensive. This may involve interviewing key stakeholders that are already identified and senior management, subject matter experts, or other project managers.

## Outputs of the Identify Stakeholders Process

The Identify Stakeholders process has two outputs: stakeholder register and stakeholder management strategy.

**Stakeholder Register** A stakeholder register captures all of the information about the stakeholders in one place. This general information includes things like the department they work in, contact information, knowledge levels, and influence levels. In addition, the stakeholder register contains at least the following:

**Identifying Information** This includes items like contact information, department, role in the project, and so on.

**Assessment Information** This includes elements regarding influence, expectations, key requirements, and when the stakeholder involvement is most critical.

**Stakeholder Classification** Stakeholders can be classified according to their relationship to the organization and, more importantly, whether they support the project, are resistant to the project, or have no opinion.

**Stakeholder Management Strategy** The stakeholder management strategy is the documented approach used to minimize negative impacts or influences that stakeholders may have throughout the life of the project. According to the *PMBOK® Guide*, the following elements should be included in the stakeholder management strategy:

- Name of key stakeholders who could have a significant impact on the project
- Stakeholders' anticipated level of participation
- Stakeholder groups
- Assessment of impact
- Potential strategies for gaining support

**Exam Essentials**

**Understand the Identify Stakeholders process.** The purpose of this process is to identify the project stakeholders, assess their influence and level of involvement, devise a plan to deal with potential negative impacts, and record stakeholder information in the stakeholder register.

# Bringing the Processes Together

As a recap, the Initiating process group kicks off the project and results in the creation of the project charter and the stakeholder register. This phase acknowledges that a new project or phase should begin.

But before a project formally exists, several scenarios may unfold:

1. First, needs and demands surface within an organization, demonstrating the need for a project to be initiated.
2. The potential project may then go through a feasibility study to determine whether the project and end result of the project are viable and whether there is an opportunity for success.
3. A project may also go through a selection process. We looked at mathematical models and benefit measurement methods as two categories of project selection methods.

After a project is selected, a project charter is created (using the Develop Project Charter process) and signed, formally authorizing the project. By the end of the Develop Project Charter process, the project charter contains several documented elements, including the justification for the project, high-level requirements and description, summary milestone schedule and budget, and the sponsor and project manager (if already selected).

Once the project charter is generated, project stakeholders can be identified. This occurs in the Identify Stakeholders process. You learned that stakeholders, along with their interests and levels of influence, are important to identify and define early on within the project. As a result, a stakeholder register, which identifies, assesses, and classifies the individual stakeholders, is created. A strategy for managing stakeholders is also developed as part of the Identify Stakeholders process.

Figure 2.14 shows the potential path of a new project, from its inception and through the initiating phase.

**FIGURE 2.14** Initiating a project

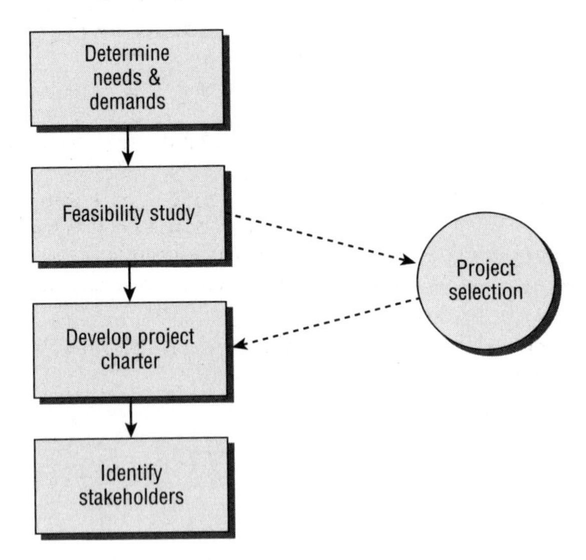


In this chapter, we also covered the nine project management knowledge areas, which are summarized in Table 2.10.

**TABLE 2.10** Project Management Knowledge Areas summary

| Knowledge Area | Description |
|---|---|
| Project Integration Management | Concerned with coordinating all the aspects of the project management plan to accomplish the project objectives |
| Project Scope Management | Concerned with managing the project scope and defining and controlling what is (and isn't) included within the project |
| Project Time Management | Concerned with completing the project on time |
| Project Cost Management | Concerned with completing the project within budget |
| Project Quality Management | Concerned with ensuring that the project satisfies the needs for which it was undertaken |
| Project Human Resource Management | Concerned with organizing, developing, and managing the project team |
| Project Communications Management | Concerned with connecting people and information together to result in successful communications throughout the project |
| Project Risk Management | Concerned with increasing the probability and impact of positive events and decreasing the probability and impact of adverse events |
| Project Procurement Management | Concerned with managing the purchasing activities of the project and the life cycle of the procurement contracts |

With the initiating phase taken care of, we are ready to move into the next phase of the project's life cycle: planning.

# Review Questions

1. Identify Stakeholders, Distribute Information, and Report Performance are all processes of which Knowledge Area?

   A. Project Integration Management

   B. Project Communications Management

   C. Project Human Resource Management

   D. Project Procurement Management

2. Ron is the project manager of a pharmaceutical company that develops multiple products to help fight diseases affecting children. There are currently two new drugs that the company is planning to develop within the next two years. Ron has been tasked with determining which of the two drugs has the greatest opportunity for success in today's marketplace. This is an example of:

   A. A business need

   B. A demand

   C. A project selection method

   D. A feasibility study

3. Your manager has recently given you the responsibility of selecting the next project, which you will manage as the project manager. There are currently three projects on hold to choose from. Using the weighted scoring models method, you determine that Project A has a weighted score of 16, Project B has a weighted score of 14, and Project C has a weighted score of 17. Which project do you choose?

   A. Project A

   B. Project B

   C. Project C

   D. None of the prospects are good selections.

4. What type of project selection method is multi-objective programming?

   A. Benefit measurement method

   B. Constrained optimization method

   C. Decision model

   D. Scoring model

5. What is $5,525 four years from now worth today given a 10 percent interest rate?

   A. $3,773.65

   B. $5,022.73

   C. $5,525.00

   D. $6,077.50

6. Which of the following BEST describes the characteristics of the product, service, or result of the project?

   A. Strategic plan

   B. Product scope description

   C. Contract

   D. Project charter

7. All of the following are inputs to the Develop Project Charter process EXCEPT:

   A. Project statement of work

   B. Business case

   C. Organizational process assets

   D. Project charter

8. Maryann has just wrapped up the final draft of the project charter and emailed a copy to the appropriate individuals. A kick-off meeting in two days has already been scheduled to complete the project charter, with all those involved having accepted the meeting invitation. What will Maryann need done during the kick-off meeting to complete the project charter?

   A. Have the charter published

   B. Confirm that the sponsor, senior management, and all key stakeholders have read and understand the charter

   C. Gather excitement and buy-in for the project among the key stakeholders

   D. Obtain sign-off of the project charter from the sponsor, senior management, and key stakeholders

9. All of the following are inputs to the Identify Stakeholders process EXCEPT:

   A. Procurement documents

   B. Project charter

   C. Project statement of work

   D. Enterprise environmental factors

10. Stakeholder management strategy can BEST be defined as:

    A. A method that captures all information about the stakeholders in one place

    B. The documented approach used to minimize negative impacts or influences that stakeholders may have throughout the life of the project

    C. A list of potential strategies for gaining the support of stakeholders

    D. A classification method that classifies all stakeholders according to their influence, expectations, key requirements, and when the stakeholder involvement is most critical

# Answers to Review Questions

1. B. Identify Stakeholders, Distribute Information, and Report Performance are all processes of the Project Communications Management Knowledge Area. Other processes of this Knowledge Area include Plan Communications and Manage Stakeholder Expectations.

2. D. A feasibility study takes place prior to making a final decision about starting a project and determines the viability of the project or product, the probability of it succeeding, and whether the technology proposed is feasible.

3. C. When projects are evaluated using the scoring model, the project with the highest overall weighted score is the best choice. Therefore, Project C, which has the highest score at 17, is the correct answer.

4. B. Multi-objective programming is a mathematical model technique. Mathematical models are also known as constrained optimization methods, making B the correct choice. As a side note, A and C refer to the same method, and D is a type of benefit measurement method.

5. A. To calculate the present value you would use the following formula:

   $PV = FV \div (1 + i)^n$

   $PV = \$5{,}525 \div (1 + 0.10)^4$

   After working out the problem, the result is a planned value of $3,773.65.

6. B. Notice that the question asks for the *best* choice. Both the product scope description and project charter provide information about the characteristics of the product, service, or result of the project, but the product scope description directly defines these characteristics, making B the best choice.

7. D. The Develop Project Charter process contains five inputs: the project statement of work, business case, contract, organizational process assets, and enterprise environmental factors. D is an output of this process.

8. D. The project charter isn't complete until sign-off has been received from the project sponsor, senior management, and key stakeholders. The signatures signify the formal authorization of the project. A project charter cannot be published until after sign-off has been obtained.

9. C. There are four inputs to the Identify Stakeholders process: procurement documents, project charter, enterprise environmental factors, and organizational process assets. C is an input to the Develop Project Charter process, and this information is reflected within the project charter input (B).

10. B. A and D both refer to the stakeholder register; C seems like a likely choice but simply refers to one part of the stakeholder management strategy. The best choice is B, which directly describes the overall purpose of the stakeholder management strategy.

Chapter 3

# Planning the Project

## THE PMP EXAM CONTENT FROM THE PLANNING PROCESS GROUP PERFORMANCE DOMAIN COVERED IN THIS CHAPTER INCLUDES THE FOLLOWING:

- ✓ Assess detailed project requirements, constraints, and assumptions with stakeholders based on the project charter, lessons learned from previous projects, and the use of requirement-gathering techniques (e.g., planning sessions, brainstorming, focus groups), in order to establish the project deliverables.
- ✓ Create the work breakdown structure with the team by deconstructing the scope, in order to manage the scope of the project.
- ✓ Develop a budget plan based on the project scope using estimating techniques, in order to manage project cost.
- ✓ Develop a project schedule based on the project timeline, scope, and resource plan, in order to manage timely completion of the project.
- ✓ Develop a human resource management plan by defining the roles and responsibilities of the project team members, in order to create an effective project organization structure and provide guidance regarding how resources will be utilized and managed.
- ✓ Develop a communication plan based on the project organization structure and external stakeholder requirements, in order to manage the flow of project information.
- ✓ Develop a procurement plan based on the project scope and schedule, in order to ensure that the required project resources will be available.

- ✓ **Develop a quality management plan based on the project scope and requirements, in order to prevent the occurrence of defects and reduce the cost of quality.**
- ✓ **Develop a change management plan by defining how changes will be handled, in order to track and manage changes.**
- ✓ **Develop a risk management plan by identifying, analyzing, and prioritizing project risks and defining risk response strategies, in order to manage uncertainty throughout the project life cycle.**
- ✓ **Present the project plan to the key stakeholders (if required), in order to obtain approval to execute the project.**
- ✓ **Conduct a kickoff meeting with all key stakeholders, in order to announce the start of the project, communicate the project milestones, and share other relevant information.**

Planning is the second of five project management process groups and includes the largest number of processes. Throughout the planning processes, the objectives are defined and refined, and a course of action is mapped out to successfully complete the project objectives as outlined by the project scope.

Several key project documents are created within this process group, including the project management plan and its collection of subsidiary plans and baselines, which will guide the project management activities.

The Planning process group accounts for 24 percent of the questions on the Project Management Professional (PMP) exam.

NOTE

The process names, inputs, tools and techniques, outputs, and descriptions of the project management process groups and related materials and figures in this chapter are based on content from *A Guide to the Project Management Body of Knowledge, 4th Edition (PMBOK® Guide).*

# Assessing the Requirements

During the early stages of planning, requirements will need to be collected, documented, and assessed. This is carried out through the Collect Requirements process (one of the 42 project management processes). This tends to be an iterative process because requirements often need to be refined as the project moves forward.

Requirements are typically conditions that must be met or criteria that the product or service of the project must possess to satisfy the objectives of the project. Requirements quantify and prioritize the wants, needs, and expectations of the project sponsor and stakeholders. According to the *PMBOK® Guide*, you must be able to measure, trace, and test requirements. It's important that they be complete and accepted by your project sponsor and key stakeholders.

As covered in Chapter 2 ("Initiating the Project") of this book, high-level requirements are documented within the project charter. Based on this document and lessons learned from past projects, detailed requirements will need to be gathered and assessed with the stakeholders. This is necessary to establishing the project deliverables. Detailed requirements can be gathered through the use of several requirement-gathering techniques that will produce a comprehensive list of requirements and that are needed to later establish the project deliverables.

Collect Requirements is the second process following the creation of the project management plan and the first process in the Project Scope Management Knowledge Area. The primary purpose of the Collect Requirements process is to define and document the project sponsors', customers', and stakeholders' expectations and needs for meeting the project objectives. This can be done through the use of lessons learned from previous projects, information from within the project charter and stakeholder register, and several requirement-gathering techniques. By the end of this process, the project will include a plan that defines how the requirements will be documented and managed throughout all phases of the project.

Figure 3.1 shows the inputs, tools and techniques, and outputs of the Collect Requirements process.

**FIGURE 3.1** Collect Requirements process

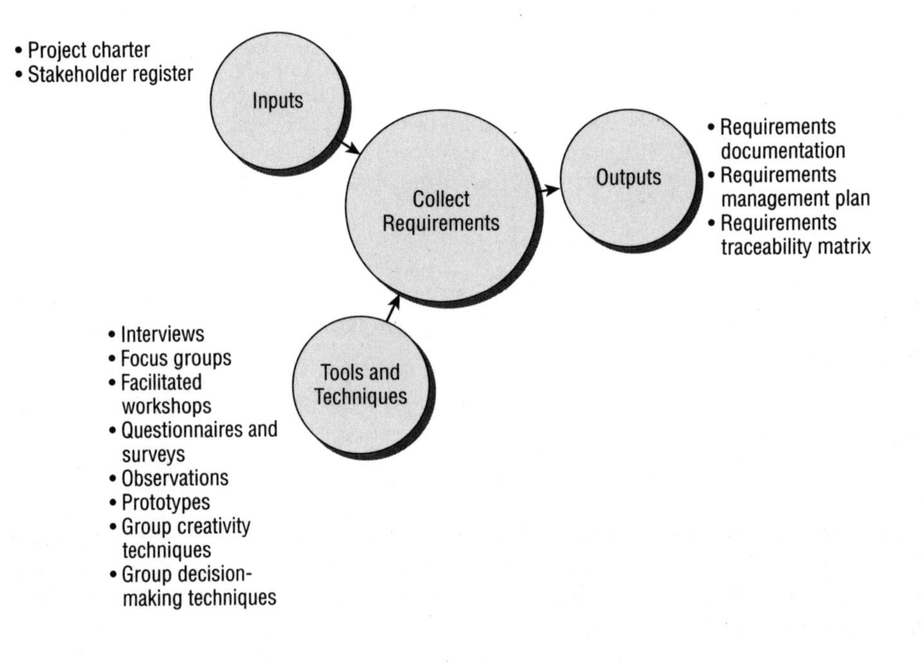

For more detailed information on the Collect Requirements process, see Chapter 3, "Developing the Project Scope Statement," of *PMP: Project Management Professional Exam Study Guide, 6th Edition* (Sybex, 2011).

## Inputs of Collect Requirements

Know the following inputs of the Collect Requirements process:

**Project Charter** In this process, the high-level project requirements and product description are used from within the project charter.

**Stakeholder Register** In this process, the stakeholder register is used as a way of determining who can provide the necessary information on the project and product requirements.

## Tools and Techniques of Collect Requirements

Know the following tools and techniques that can be used during the Collect Requirements process:

- Interviews
- Focus groups
- Facilitated workshops
- Questionnaires and surveys
- Observations
- Prototypes
- Group creativity techniques
- Group decision-making techniques

**Interviews** Interviews allow subject matter experts and experienced project participants to impart a lot of information in a short amount of time. The following are characteristics of interviews:

- Typically structured as a one-on-one conversation with stakeholders
- Can be formal or informal
- Consist of questions prepared ahead of time

**Focus Groups** Focus groups are usually conducted by a trained moderator. The key to this tool lies in picking the subject matter experts and stakeholders to participate in the focus group; they then engage in discussion and conversation as a way of imparting their feedback about the project's end result.

**Facilitated Workshops** Cross-functional stakeholders come together in a facilitated workshop to discuss and define requirements that affect more than one department. The following are characteristics of facilitated workshops:

- All the participants of the workshops understand the various project needs.
- The workshops provide a facilitated forum to discuss and resolve their participants' issues.
- A consensus is reached more easily than through other methods.

The primary difference between focus groups and facilitated workshops is that focus groups are gatherings of prequalified subject matter experts and stakeholders where the intention is to gather feedback from these individuals, and facilitated workshops consist of cross-functional stakeholders who work together to define cross-functional requirements.

**Questionnaires and Surveys** Questionnaires and surveys involve querying a large group of participants via questionnaires or surveys and applying statistical analysis, if needed, to the results. This is considered to be a quick way of obtaining feedback from a large number of stakeholders.

**Observations** Observations consist of a one-on-one experience where an observer sits side by side with the participant to observe how the participant interacts with the product or service. This technique is also known as job shadowing, and it can also involve participant observers who perform the job themselves to ascertain requirements. Observations are particularly useful when stakeholders have a difficult time verbalizing requirements.

**Prototypes** Prototyping is a technique involving constructing a working model or mock-up of the final product for participants to experiment with. The prototype does not usually contain all the functionality the end product does, but it gives participants enough information that they can provide feedback regarding the mock-up. This is an iterative process where participants experiment and provide feedback, the prototype is revised, and the cycle starts again. Prototypes are a great way of getting feedback during the early stages of a project's life cycle.

**Group Creativity Techniques** Group creativity involves the following techniques:

**Brainstorming** Brainstorming involves getting together to generate ideas and gather the project requirements.

**Nominal Group Technique** Nominal group technique works to engage all participants through an idea-gathering or structured brainstorming session and ranking process.

**Delphi Technique** The Delphi technique utilizes a group of expert responses and feedback. Experts answer questionnaires by providing feedback on responses from each round of requirement gathering while maintaining a level of anonymity, thereby reducing bias in their feedback.

**Idea/Mind Mapping** In idea/mind mapping, participants use a combination of brainstorming and diagramming techniques to record their ideas.

**Affinity Diagrams** Affinity diagrams sort ideas into related groups for further analysis and review.

**Group Decision-Making Techniques** According to the *PMBOK® Guide*, several group decision-making techniques are used. The four methods mentioned are as follows:

- Unanimity, where everyone agrees on the resolution or course of action
- Majority, where more than 50 percent of the members support the resolution

- Plurality, where the largest subgroup within the group makes the decision if a majority is not reached
- Dictatorship, where one person makes the decision on behalf of the group

## Outputs of Collect Requirements

For the exam, know the outputs of the Collect Requirements process, which are as follows:

**Requirements Documentation** This output involves recording the requirements in a requirements document. According to the *PMBOK® Guide*, this document includes at least the following elements:

- Business need for the project and why it was undertaken
- Objectives of the project and the business objectives the project hopes to fulfill
- Functional requirements
- Nonfunctional requirements
- Quality requirements
- Acceptance criteria
- Business rules
- Organizational areas and outside entities impacted
- Support and training requirements
- Assumptions and constraints
- Signatures of the key stakeholders

**Requirements Management Plan** The requirements management plan defines how the requirements will be analyzed, documented, and managed throughout all phases of the project. The type of phase relationship used to manage the project will determine how requirements are managed throughout the project. A requirements management plan includes the following components:

- How planning, tracking, and reporting of requirements activities will occur
- How changes to the requirements will be requested, tracked, and analyzed along with other configuration management activities
- How requirements will be prioritized
- Which metrics will be used to trace product requirements
- Which requirements attributes will be documented in the traceability matrix

**Requirements Traceability Matrix** The traceability matrix links requirements to business needs and project objectives and also documents the following:

- Where the requirement originated
- What the requirement will be traced to
- Status updates all the way through delivery or completion

According to the *PMBOK® Guide*, the requirements traceability matrix helps assure that business value is realized when the project is complete because each requirement is linked to a business and project objective.

Table 3.1 shows a sample traceability matrix with several attributes that identify the requirement.

**TABLE 3.1** Requirements traceability matrix

| Unique ID | Description of Requirement | Source | Priority | Test Scenario | Test Verification | Status |
|---|---|---|---|---|---|---|
| 001 | Requirement one | Project objective and stakeholder | B | User acceptance | | Approved |

Here are the characteristics of a traceability matrix:

- Each requirement should have its own unique identifier.
- A brief description includes enough information to easily identify the requirement.
- The source column refers to where the requirement originated.
- Priority refers to the priority of the requirement.
- The test scenario records how the requirement will be tested or during which project phase, and test verification indicates if it passes or fails the test.
- Status may capture information about the requirement that refers to whether the requirement has been approved to be included in the project. Examples include added, deferred, and canceled.

## Exam Essentials

**Understand the purpose of collecting requirements.** Requirements are collected to define and document the project sponsors', customers', and stakeholders' expectations and needs for meeting the project objective.

# Creating the Work Breakdown Structure

After requirements have been gathered, the next step is to begin breaking down and documenting the project and product's scope. The end result will be a work breakdown structure (WBS), which is a deliverable-oriented, high-level decomposition of the project work. But before the WBS can be generated, there are a few additional steps that must be addressed:

- The creation of a scope management plan
- The creation and sign-off of the project scope statement
- The creation of a scope baseline

Although the scope management plan isn't a formal output of a process, it is an important document that exists prior to the start of any scope-related process. The project scope statement is an output of the Define Scope process and is where the project deliverables are documented along with the constraints and assumptions of the project. Once the project scope statement, WBS, and another output called the WBS dictionary are created, they become part of a scope baseline.

## Understand the Scope Management Plan

As mentioned previously, the scope management plan is not an official output of a process, but it plays an important role within all of the Project Scope Management Knowledge Area processes. Plans not created out of a formal process are considered to be generated out of the Develop Project Management Plan process, which is high-level planning process that will be covered later in this chapter.

The scope management plan contains the following information:

- The process used to prepare the project scope statement
- A process for creating the work breakdown structure (WBS)
- A definition of how the deliverables will be verified for accuracy and the process used for accepting deliverables
- A description of the process for controlling scope change requests, including the procedure for requesting changes and how to obtain a change request form

Keep in mind that throughout the Project Scope Management Knowledge Area, the scope management plan serves as a guide for documenting and controlling the scope of the project.

## Define Scope

The result and main objective of the Define Scope process is the creation of the project scope statement. The project scope statement is used to develop and document a detailed description of the deliverables of the project and the work needed to produce them. This process addresses the project objectives, requirements, constraints, assumptions, and other elements of writing the project scope statement and is progressively elaborated as more detail becomes known.

Figure 3.2 shows the inputs, tools and techniques, and outputs of the Define Scope process.

**FIGURE 3.2** Define Scope process

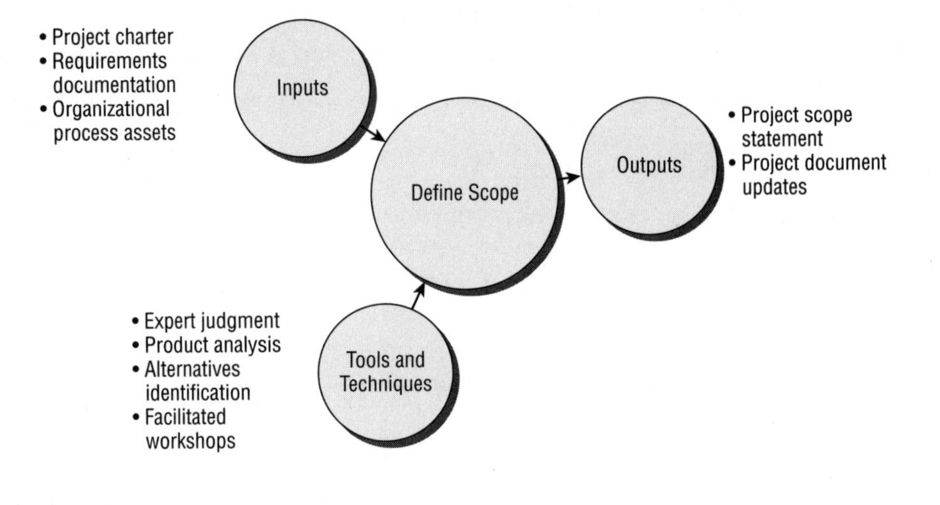

For more detailed information on the Define Scope process, see Chapter 3 of *PMP: Project Management Professional Exam Study Guide, 6th Edition*.

## Inputs of Define Scope

The Define Scope process has three inputs you should know:

**Project Charter** From within the project charter, this process primarily utilizes the project objectives. Objectives are quantifiable criteria used to measure project success. They describe *what* the project is trying to do, accomplish, or produce. Quantifiable criteria should at least include the following items:

- Schedule
- Cost
- Quality measures
- Business measures or quality targets (sometimes)

**Requirements Documentation** The requirements documentation includes elements such as functional and nonfunctional characteristics, quality metrics, and acceptance criteria.

For a more detailed description, see “Outputs of Collect Requirements” earlier in this chapter.

**Organizational Process Assets** This process utilizes historical information from within the organizational process assets as well as policies, procedures, and project scope statement templates.

## Tools and Techniques of Define Scope

Know the following tools and techniques that you will use during the Define Scope process:

**Expert Judgment** According to the *PMBOK® Guide*, the following items are examples of expert judgment utilized within this process:

- Subject matter experts or consultants
- Stakeholders
- Industry groups and associations
- Other departments or units internal to the organization

**Product Analysis** Product analysis is a method for converting the product description and project objectives into deliverables and requirements. According to the *PMBOK® Guide*, product analysis might include performing the following to further define the product or service:

- Value analysis
- Functional analysis
- Requirements analysis
- Systems-engineering techniques
- Systems analysis
- Product breakdown
- Value-engineering techniques

**Alternatives Identification** Alternatives identification is a technique used for discovering different methods or ways of accomplishing the work of the project. Examples of alternatives identification are brainstorming and lateral thinking.

Lateral thinking is a process of separating problems, looking at them from angles other than their obvious presentation, and encouraging team members to come up with ways to solve problems that are not apparently obvious or look at scope in a different way. This can also be described as out-of-the-box thinking.

**Facilitated Workshops** Facilitated workshops involve stakeholders coming together to discuss and define requirements that affect more than one department.

## Outputs of Define Scope

Know the following outputs of the Define Scope process:

**Project Scope Statement** The purpose of the project scope statement is to document the project objectives, deliverables, and the work required to produce the deliverables so that it can be used to direct the project team's work and as a basis for future project decisions. The project scope statement is an agreement between the project organization and the customer that states precisely what the work of the project will produce.

The project scope statement guides the work of the project team during the Executing process, and all change requests will be evaluated against the scope statement. The criteria outlined will also be used to determine whether the project was completed successfully.

According to the *PMBOK® Guide*, the project scope statement should include all of the following:

- Product scope description
- Product acceptance criteria
- Project deliverables
- Project exclusions
- Project constraints
- Project assumptions

**Product Scope Description** The product scope description describes the characteristics of the product, service, or result of the project.

**Product Acceptance Criteria** Product acceptance criteria include the process and criteria that will be used to determine whether the deliverables and the final product, service, or result of the project is acceptable and satisfactory.

**Project Deliverables** Deliverables are measurable outcomes, measurable results, or specific items that must be produced to consider the project or project phase completed. Deliverables should be specific, unique, and verifiable and may include supplementary outcomes.

### Deliverables vs. Requirements

Deliverables describe the components of the goals and objectives in a quantifiable way. Requirements are the specifications of the deliverables. Select deliverables and requirements are sometimes referred to as *critical success factors.* Critical success factors are those elements that must be completed for the project to be considered complete.

**Project Exclusions** Project exclusions are anything that isn't included as a deliverable or work of the project. Project exclusions should be noted in the project scope statement for stakeholder management purposes.

**Project Constraints** Anything that either restricts the actions of the project team or dictates the actions of the project team or the way the project should be performed is considered a constraint. The project manager must balance the project constraints while meeting or exceeding the expectations of the stakeholders.

Examples of project constraints include, but are not limited to, the following items:

- Scope
- Quality
- Schedule
- Budget
- Resources
- Risk

**Project Assumptions** Assumptions are things considered true, real, or certain, for planning purposes. Each project will have its own set of assumptions, and the assumptions should be identified, documented, validated, and updated throughout the project. Defining new assumptions and refining old ones are forms of progressive elaboration. The following list includes examples of assumptions:

- Vendor delivery times
- Product availability
- Contractor availability
- Accuracy of the project plan
- Belief that key project members will perform adequately
- Contract signing dates
- Project start dates
- Project phase start dates

**Project Document Updates** Project documents encompass those documents from within the project that are not plans or baselines. There may be several common updates and changes to project documents resulting from this process:

- Original project objectives
- Stakeholder register
- Requirements documentation
- Requirements traceability matrix
- The scope statement as a result of approved changes

In practice, the Define Scope process is performed before the Collect Requirements process. Deliverables must be identified before their detailed requirements are documented.

## Create WBS

The Create WBS process takes the well-defined deliverables and requirements and begins the process of breaking down the work via a WBS. WBS stands for work breakdown structure, which defines the scope of the project and breaks down the deliverables into smaller, more manageable components of work that can be scheduled and estimated as well as easily assigned, monitored, and controlled. The WBS should detail the full scope of work needed to complete the project.

Figure 3.3 shows the inputs, tools and techniques, and outputs of the Create WBS process.

**FIGURE 3.3** Create WBS process

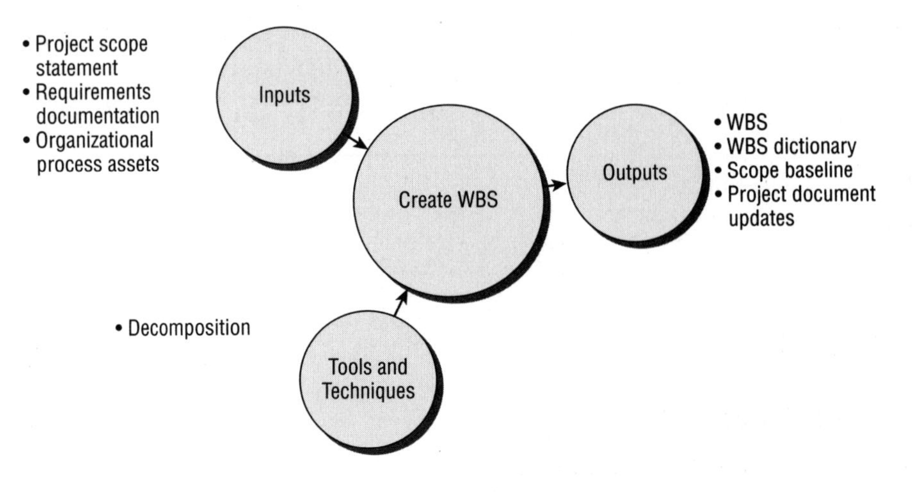


### Inputs of Create WBS

For the exam, know the following inputs of the Create WBS process:

**Project Scope Statement** The project scope statement contains information valuable to creating the WBS, most notably the list of project deliverables.

**Requirements Documentation** The requirements documentation describes how the requirements meet the business needs of the project.

**Organizational Process Assets** This process utilizes historical information from within the organizational process assets as well as policies, procedures, and WBS templates.

For more detailed information on the Create WBS process, see Chapter 3 of *PMP: Project Management Professional Exam Study Guide, 6th Edition.*

## Tools and Techniques of Create WBS

The Create WBS process has only one tool and technique: decomposition.

Decomposition involves breaking down the project deliverables into smaller, more manageable components of work that can be easily planned, executed, monitored and controlled, and closed out.

The project manager typically decomposes the work with the help of the project team, whose members have expert knowledge of the work. Benefits of team involvement include a more realistic decomposition of the work and getting team buy-in.

Decomposition provides a way of managing the scope of the project and also does the following:

- Improves estimates
- More easily assigns performance measures and controls
- Provides a baseline to compare against throughout the project or phase
- Ensures that assignments go to the proper parties

According to the *PMBOK® Guide*, decomposition consists of the following five-step process:

1. Identify the deliverables and work.
2. Organize the WBS.
3. Decompose the WBS components into lower-level components.
4. Assign identification codes.
5. Verify the WBS.

## Outputs of Create WBS

The Create WBS process has four outputs:

- WBS
- WBS dictionary
- Scope baseline
- Project document updates

**Work Breakdown Structure (WBS)** The following are various ways of organizing the WBS:

**Major Deliverables** The major deliverables of the project are used as the first level of decomposition in this structure.

**Subprojects** Another way to organize the work is by subprojects. Each of the subproject managers will develop a WBS for their subproject that details the work required for that deliverable.

**Project Phases** Many projects are structured or organized by project phases. Each phase listed here would be the first level of decomposition, and its deliverables would be the next level.

The following is a general description of the various levels within the WBS:

**Level 1** According to the *PMBOK® Guide*, Level 1 of the WBS is the project level. The first level of decomposition, however, may be the second level of the WBS, which could include deliverables, phases, and subprojects—see Figure 3.4 for an example.

**FIGURE 3.4** WBS Level 1 and Level 2

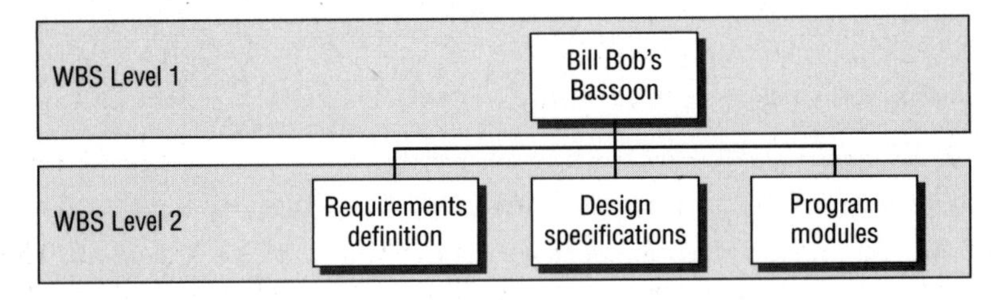


**Level 2** Level 2 and levels that follow show more and more detail. Each of these breakouts is called a *level* in the WBS. See Figures 3.5 and 3.6 for an example.

**FIGURE 3.5** WBS Levels 1 through 3

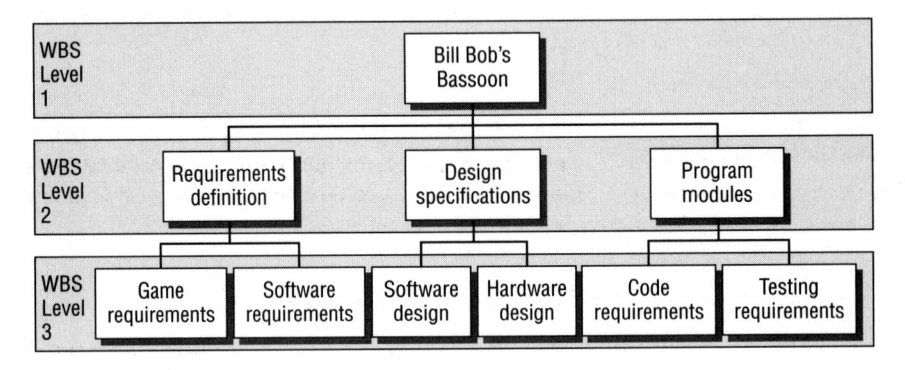

**FIGURE 3.6** WBS Levels 1 through 4

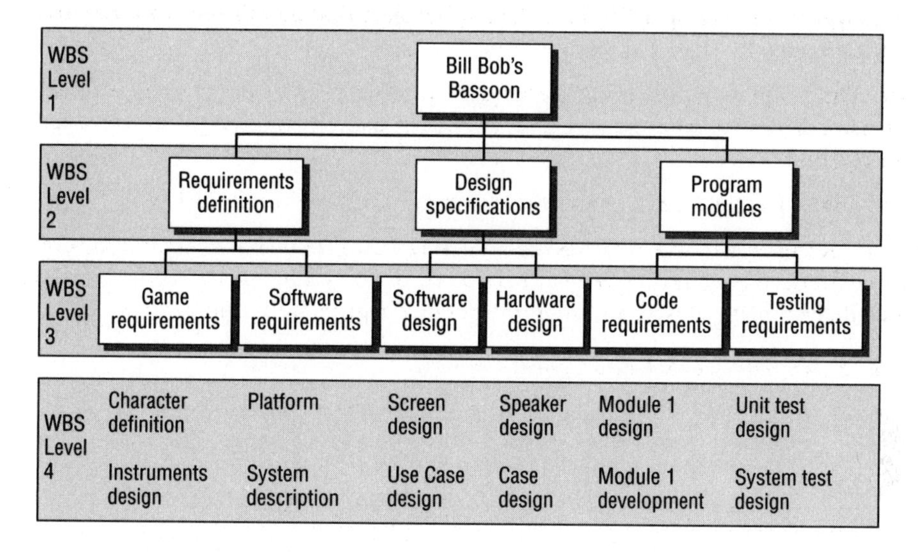


**Lowest Level** The lowest level of any WBS is called the work package level. The goal is to construct the WBS to the work package level where cost estimates and schedule dates can be estimated reliably and easily.

NOTE Work packages are the components that can be easily assigned to one person, or a team of people, with clear accountability and responsibility for completing the assignment. The work package level will later be decomposed further into lists of activities.

Collectively, all the levels of the WBS roll up to the top so that all the work of the project is captured. According to the *PMBOK® Guide*, this is known as the 100 percent rule.

Here are the other elements of the WBS that you should know:

**WBS Templates** Work breakdown structures can be constructed using WBS templates or the WBS from a similar completed project. Although every project is unique, many companies and industries perform the same kind of projects repeatedly.

**Rolling Wave Planning** Rolling wave planning is a process of elaborating deliverables, project phases, or subprojects in the WBS to differing levels of decomposition depending on the expected date of the work.

**Unique WBS Identifiers** Each element at each level of the WBS is generally assigned a unique identifier according to the *PMBOK® Guide*. Figure 3.7 shows what a WBS with unique identifiers displayed might look like.

**FIGURE 3.7** Unique WBS identifiers

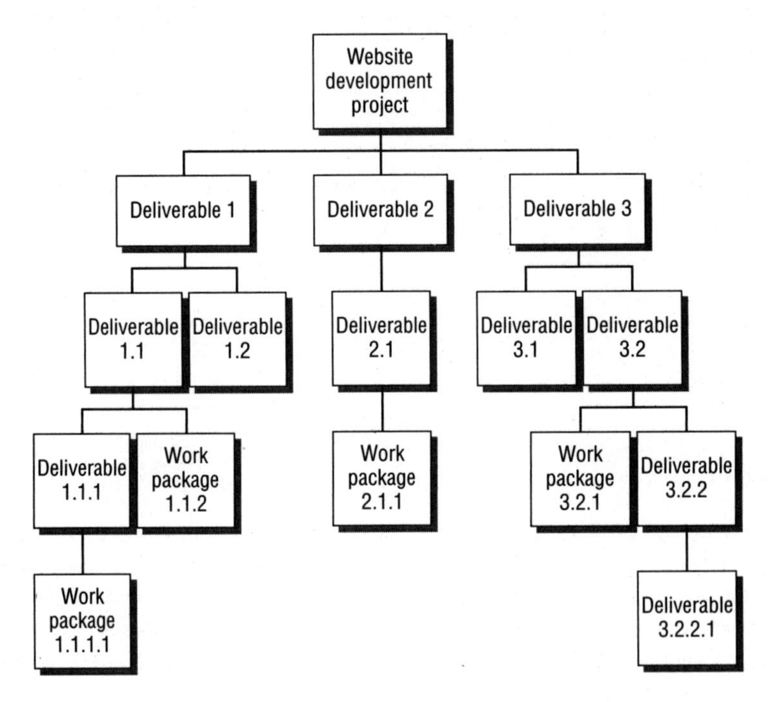


**WBS Dictionary** The WBS dictionary is where work component descriptions are documented. According to the *PMBOK® Guide*, the WBS dictionary should include the following elements for each component of the WBS:

- Code of accounts identifier
- Description of the work of the component
- Organization responsible for completing the component
- List of schedule milestones
- Schedule activities associated with the schedule milestones
- Required resources
- Cost estimates
- Quality requirements
- Criteria for acceptance

- Technical references
- Contract information

**Scope Baseline** The scope baseline for the project is the approved project scope statement, the WBS, and the WBS dictionary. These documents together describe in detail all the work of the project. These documents allow managers to carry out the following activities:

- Document schedules
- Assign resources
- Monitor and control the project work

**Project Document Updates** The following documents may need to be updated as a result of this process:

- Project scope statement
- Project management plan
- Any other project documents that should reflect approved changes to the project scope statement

## Exam Essentials

**Understand the purpose of the project scope statement.** The scope statement serves as a common understanding of project scope among the stakeholders. The project objectives and deliverables and their quantifiable criteria are documented in the scope statement and are used by the project manager and the stakeholders to determine whether the project was completed successfully. It also serves as a basis for future project decisions.

**Be able to define project constraints and assumptions.** Project constraints limit the options of the project team and restrict their actions. Sometimes, constraints dictate actions. Common constraints include scope, quality, schedule, budget, resources, and risk. Assumptions are conditions that, for planning purposes, are presumed to be true, real, or certain.

**Be able to describe the purpose of the scope management plan.** The scope management plan has a direct influence on the project's success and describes the process for determining project scope, facilitates creating the WBS, describes how the product or service of the project is verified and accepted, and documents how changes to scope will be handled. The scope management plan is a subsidiary plan of the project management plan.

**Be able to define a WBS and its components.** The WBS is a deliverable-oriented hierarchy. It uses the deliverables from the project scope statement or similar documents and decomposes them into logical, manageable units of work. The lowest level of any WBS is called a work package.

# Developing a Cost Management Plan

The purpose of the cost management plan is, naturally, to create, monitor, and control the project costs. This plan is based on the project scope and documents the estimating techniques that will be used as well as how the cost-related processes will be carried out. Like the scope management plan, the cost management plan is not a formal output of a process but should be created prior to beginning the Project Cost Management Knowledge Area processes.

The project budget is created by carrying out two planning processes:

- Estimate Costs, which estimates how much each activity will cost
- Determine Budget, which aggregates the total cost estimates plus contingency reserves to create the project budget

The project budget is referred to as the cost performance baseline, which, along with the cost management plan, becomes a part of the project management plan.

## Understand the Cost Management Plan

Although not an official output of a process, the cost management plan plays an important role within all of the Project Cost Management Knowledge Area processes. The cost management plan is created during the Develop Project Management Plan process and is a subsidiary plan of the project management plan (as all subplans are). The cost management plan contains the following information:

- Level of accuracy
- Units of measure
- Organizational procedures links
- Control thresholds
- Rules of performance measurement
- Reporting formats
- Process descriptions

Using the preceding components, the cost management plan will guide the project management team in carrying out the three cost-related processes.

## Estimate Costs

The purpose of the Estimate Costs process is to develop cost estimates for resources, both human and material, required for all schedule activities and the overall project. This includes weighing alternative options and examining risks and trade-offs. The cost-related processes are governed by the cost management plan, which establishes the format and conditions used to plan for project costs. It also outlines how you will estimate, budget, and control project costs.

Figure 3.8 shows the inputs, tools and techniques, and outputs of the Estimate Costs process.

**FIGURE 3.8** Estimate Costs process

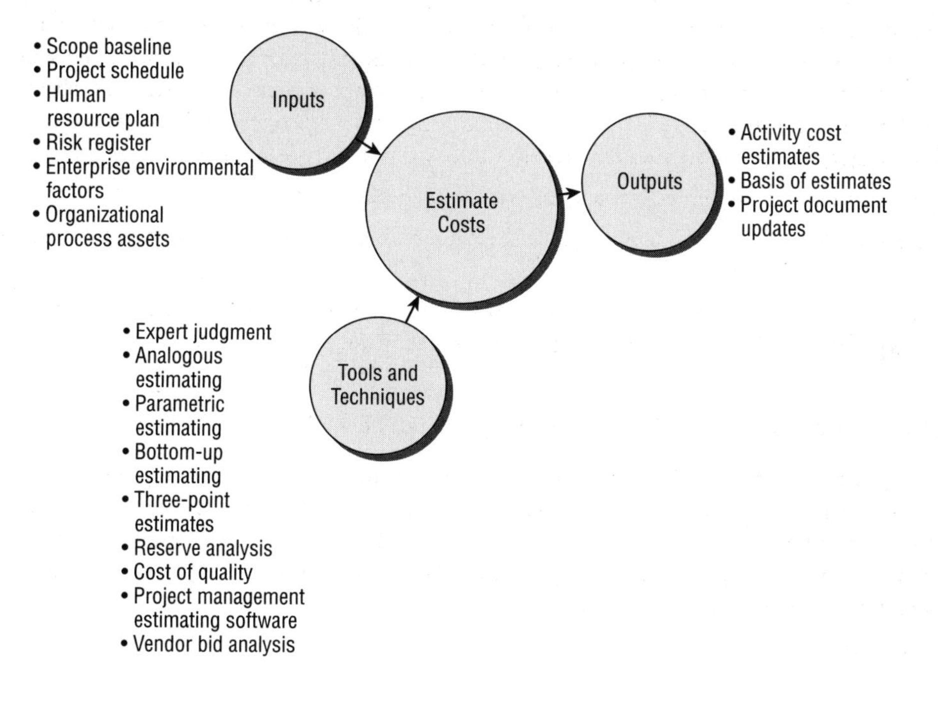

For more detailed information on the Estimate Costs process, see Chapter 5, "Developing the Project Budget," of *PMP: Project Management Professional Exam Study Guide, 6th Edition.*

## Inputs of Estimate Costs

There are six inputs of the Estimate Costs process that you should be familiar with:

- Scope baseline
- Project schedule
- Human resource plan
- Risk register
- Enterprise environmental factors
- Organizational process assets

**Scope Baseline** The following are included within the scope baseline and utilized in creating cost estimates:

- Project scope statement (key deliverables, constraints, and assumptions)
- WBS, which serves as the basis for estimating costs (project deliverables, control accounts)
- WBS dictionary

Scope definition is a key component of determining the estimated costs and should be completed early within the project.

**Project Schedule** Activity resource requirements and activity duration estimates are the key outputs that should be considered when estimating costs.

**Human Resource Plan** For the purposes of the Estimate Costs process, the following elements of the human resource plan should be considered:

- Personnel rates
- Project staffing attributes
- Employee recognition or rewards programs

**Risk Register** From within the risk register, the cost of mitigating risks will be utilized for estimating costs, particularly the risks with negative impacts to the project.

**Enterprise Environmental Factors** According to the *PMBOK® Guide*, the following enterprise environmental factors should be considered in this process:

- Market conditions, which help you to understand the materials, goods, and services available in the market and what terms and conditions exist to procure those resources.
- Published commercial information, which refers to resource cost rates. These are obtained from commercial databases or published seller price lists.

**Organizational Process Assets** The organizational process assets considered include historical information and lessons learned on previous projects of similar scope and complexity. Also useful are cost-estimating worksheets from past projects as templates for the current project.

## Tools and Techniques of Estimate Costs

The following list includes the tools and techniques of the Estimate Costs process:

- Expert judgment
- Analogous estimating
- Parametric estimating

- Bottom-up estimating
- Three-point estimates
- Reserve analysis
- Cost of quality
- Project management estimating software
- Vendor bid analysis

**Expert Judgment** Cost estimates from individuals who have previous experience in past similar projects can be utilized. Information from those who have experience in performing the activities becomes valuable for calculating accurate estimates.

**Analogous Estimating** Analogous estimating, also called top-down estimating, is a form of expert judgment. This technique uses actual costs of a similar activity completed on a previous project to determine the cost estimate of the current activity. It can be used when the previous activities being compared are similar to the activities being estimated. Analogous estimating may also be useful when detailed information about the project is not yet available. Top-down estimating techniques are also used to estimate total project cost to look at the estimate as a whole. While analogous estimating is considered to be a quick and low-cost estimating technique, it is low in accuracy.

The *PMBOK® Guide* states that analogous estimating can also be used to determine overall project duration and cost estimates for the entire project (or phases of the project).

**Parametric Estimating** Parametric estimating uses historical information and statistical data to calculate cost estimates. For a complete description of parametric estimating, see "Tools and Techniques of Estimate Activity Durations" later in this chapter.

**Bottom-Up Estimating** This technique estimates costs associated with every activity individually by decomposing that activity into smaller pieces of work until the cost of the work can be confidently estimated. The costs of the activity are then rolled up to the original activity level.

**Three-Point Estimates** Three-point estimates are used in this process when the project team is attempting to improve estimates and account for risk and estimate uncertainty.

**Reserve Analysis** During this process, cost reserves (or contingencies) are added. Cost contingencies can be aggregated and assigned to a schedule activity or a WBS work package level. This reserve is added to account for existing risk.

**Cost of Quality** The cost of quality (COQ) is the total cost to produce the product or service of the project according to the quality standards. You can find additional information on cost of quality in "Tools and Techniques of Plan Quality" later in this chapter.

**Project Management Estimating Software** The project management estimating software tool can help quickly determine estimates given different variables and alternatives.

**Vendor Bid Analysis** Vendor bid analysis involves gathering information from vendors to help establish cost estimates. This can be accomplished by requesting bids or quotes or working with some of the trusted vendor sources for estimates.

## Outputs of Estimate Costs

The following are three outputs that result from the Estimate Costs process:

- Activity cost estimates
- Basis of estimates
- Project document updates

**Activity Cost Estimates** Activity cost estimates are quantitative amounts that reflect the cost of the resources needed to complete the project activities. The following resources are commonly needed:

- Human resources
- Material
- Equipment
- Information technology needs
- Any contingency reserve amounts and inflation factors

**Basis of Estimates** Basis of estimates is the supporting detail for the activity cost estimates and includes any information that describes how the estimates were developed, what assumptions were made during the Estimate Costs process, and any other details needed. According to the *PMBOK® Guide*, the basis of estimates should include the following as a minimum:

- Description of how the estimate was developed or the basis for the estimate
- Description of the assumptions made about the estimates or the method used to determine them
- Description of the constraints
- Range of possible results
- Confidence level regarding the final estimates

**Project Document Updates** Updates to project documents, as a result of determining cost estimates, commonly include the following to account for cost variances and refined estimates:

- Project budget
- Risk register

## Determine Budget

The Determine Budget process is primarily concerned with determining the cost performance baseline, which represents the project budget. This process aggregates the cost estimates of activities and establishes a cost performance baseline for the project that is used to measure performance of the project throughout the remaining process groups. By the end of this process, the project schedule and the project budget will both have been created.

Figure 3.9 shows the inputs, tools and techniques, and outputs of the Determine Budget process.

**FIGURE 3.9** Determine Budget process

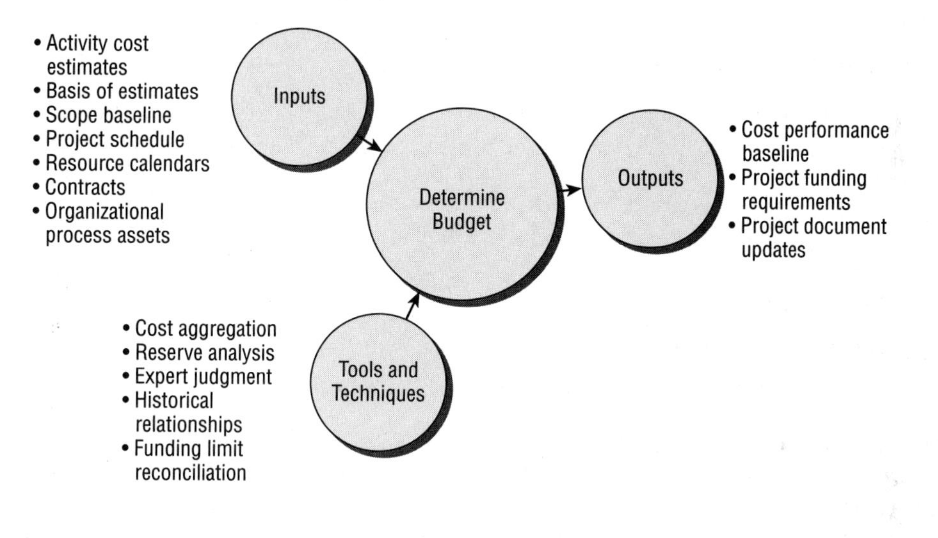

For more detailed information on the Determine Budget process, see Chapter 5 of *PMP: Project Management Professional Exam Study Guide, 6th Edition*.

### Inputs of Determine Budget

The Determine Budget process contains seven inputs:

- Activity cost estimates
- Basis of estimates
- Scope baseline
- Project schedule
- Resource calendars
- Contracts
- Organizational process assets

**Activity Cost Estimates** Activity cost estimates are determined for each activity within a work package and then summed to determine the total estimate for a work package. This information is essential to creating a project budget.

**Basis of Estimates** Basis of estimates contains all the supporting detail regarding the estimates. Assumptions regarding indirect costs and whether they will be included in the project budget should be considered. Indirect costs cannot be directly linked to any one project but are allocated among several projects.

**Scope Baseline** When determining the budget, the following is utilized within the scope baseline:

- Project scope statement (describes the constraints of the project)
- WBS (shows how the project deliverables are related to their components and typically provides control account information)
- WBS dictionary

**Project Schedule** The schedule contains information that is helpful in developing the budget, such as start and end dates for activities, milestones, and so on. Based on the information in the schedule, budget expenditures for calendar periods can be determined.

**Resource Calendars** Resource calendars help determine costs in calendar periods and over the length of the project because they describe what resources are needed for the project.

**Contracts** Contracts include cost information that should be included in the overall project budget.

**Organizational Process Assets** The organizational process assets utilized include the following items:

- Cost budgeting tools
- Policies and procedures of the organization regarding budgeting exercises
- Reporting methods

Outputs from other planning processes, including Create WBS, Develop Schedule, and Estimate Costs, must be completed prior to working on Determine Budget because some of their outputs become the inputs to this process.

## Tools and Techniques of Determine Budget

Know the following tools and techniques of the Determine Budget process for the exam:

- Cost aggregation
- Reserve analysis
- Expert judgment

- Historical relationships
- Funding limit reconciliation

**Cost Aggregation** Cost aggregation is the process of tallying the schedule activity cost estimates at the work package level and then totaling the work package levels to higher-level WBS component levels. All costs can then be aggregated to obtain a total project cost.

**Reserve Analysis** Within this process, reserve analysis plans contingency reserves for unplanned project scope and project costs. Contingency reserves are included as part of the cost performance baseline, while management reserves are not.

**Expert Judgment** Expert judgment in calculating the project budget will include information from those with past experience in conducting similar projects.

**Historical Relationships** Analogous estimates and parametric estimates can be used to help determine total project costs. Actual costs from previous projects of similar size, scope, and complexity are used to estimate the costs for the current project. This is helpful when detailed information about the project is not available or it's early in the project phases and not much information is known.

**Funding Limit Reconciliation** Funding limit reconciliation involves reconciling the amount of funds to be spent with the amount of funds budgeted for the project. The organization or the customer sets these limits. Reconciling the project expenses will require adjusting the schedule so that the expenses can be smoothed. This can be done by placing imposed date constraints on work packages or other WBS components in the project schedule.

## Outputs of Determine Budget

There are three outputs of the Determine Budget process:

**Cost Performance Baseline** The cost performance baseline provides the basis for measurement, over time, of the expected cash flows (or funding disbursements) against the requirements. Adding the costs of the WBS elements by time periods develops the cost performance baseline. This is also known as the project's time-phased budget. Most projects span some length of time, and most organizations time the funding with the project.

Cost performance baselines can be displayed graphically, as shown in Figure 3.10.

The cost performance baseline should contain the costs for all of the expected work on the project, as well as the contingency reserves set aside to account for known risks. Large projects may have more than one cost performance baseline.

Cost performance baselines are displayed as an S curve. The reason for this is that project spending starts out slowly, gradually increases over the project's life until it reaches a peak, and then tapers off again as the project wraps up.

**FIGURE 3.10** Cost performance baseline

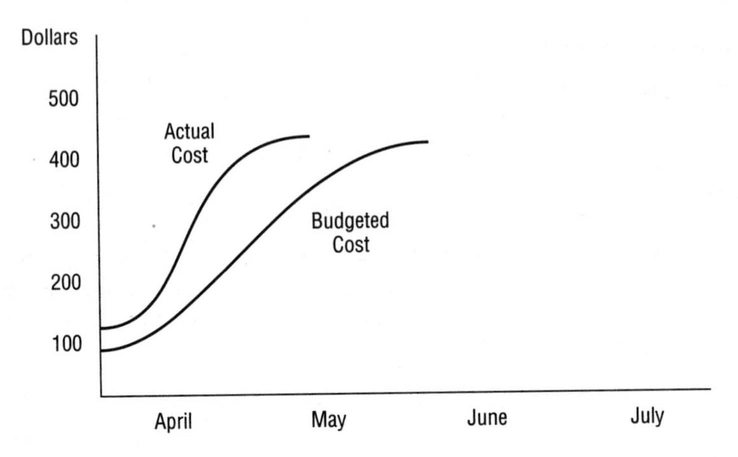


**Project Funding Requirements** Project funding requirements describe the need for funding over the course of the project and are derived from the cost performance baseline. Funding requirements can be expressed in monthly, quarterly, or annual increments or other increments that are appropriate for your project.

Figure 3.11 shows the cost performance baseline, the funding requirements, and the expected cash flows plotted on the S curve. This example shows a negative amount of management reserve. The difference between the funding requirements and the cost performance baseline at the end of a project is the management reserve. Management reserves are set aside to deal with unknown-unknown risks.

**FIGURE 3.11** Cost performance baseline, funding requirements, and cash flow

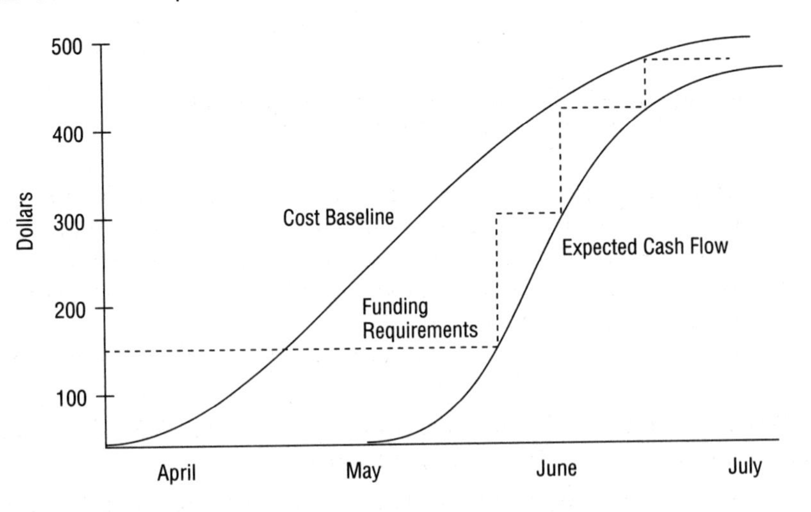

**Project Document Updates** The following project documents require updates as a result of carrying out this process:

- Cost management plan
- Any other project documents that include cost estimates

**Exam Essentials**

**Be able to identify and describe the primary output of the Estimate Costs process.** The primary output of Estimate Costs is activity cost estimates. These estimates are quantitative amounts—usually stated in monetary units—that reflect the cost of the resources needed to complete the project activities.

**Be able to identify the tools and techniques of the Estimate Costs process.** The tools and techniques of Estimate Costs are expert judgment, analogous estimating, parametric estimating, bottom-up estimating, three-point estimate, reserve analysis, cost of quality, project management estimating software, and vendor bid analysis.

**Be able to identify the tools and techniques of the Determine Budget process.** The tools and techniques of Determine Budget are cost aggregation, reserve analysis, expert judgment, historical relationships, and funding limit reconciliation.

**Be able to describe the cost performance baseline.** The cost performance baseline is the authorized, time-phased cost of the project when using budget-at-completion calculations. The cost performance baseline is displayed as an S curve.

**Be able to describe project funding requirements.** Project funding requirements are an output of the Determine Budget process. They detail the funding requirements needed on the project by time period (monthly, quarterly, annually).

# Developing a Project Schedule

Like the Project Scope and Project Cost Management Knowledge Areas, the Project Time Management Knowledge Area also has a plan that guides the processes within it: the schedule management plan. Similar to the scope and cost management plans, it is not a formal output of a process, but it plays an important role in guiding the project management team in carrying out the six time-related processes.

Developing a schedule consists of five steps, each carried out through a formal planning process:

- Define Activities, which breaks down the work packages of the WBS into detailed work (activities) to be carried out by project team members
- Sequence Activities, which places the activities into a logical order based on existing dependencies

- Estimate Activity Resources, which estimates the type and quantity of resources needed to carry out each activity
- Estimate Activity Durations, which estimates how long each activity will take
- Develop Schedule, which produces an accepted and signed-off project schedule

Through the use of the planning processes noted, the project management team builds a schedule that is based on the project's timeline, scope, and resource plan to arrive at the timely completion of the project. The accepted and signed-off version of the project schedule becomes the schedule baseline.

## Understand Schedule Management Plan

The schedule management plan should be created prior to beginning any of the time-related processes and early within the planning stages of a project. The plan is responsible for guiding the creation, management, and control of the project schedule.

According to the *PMBOK® Guide*, this plan consists of the following elements:

- Scheduling methodology
- Scheduling tool
- Schedule format
- Criteria for developing and controlling the project schedule

The Develop Project Management Plan process is responsible for any plan that is not formally created as an output of another process. This is most notably the case for the scope, cost, and schedule management plans. The Develop Project Management Plan process is the second process of the Project Integration Management Knowledge Area and is responsible for creating the project management plan.

## Define Activities

The purpose of the Define Activities process is to decompose the work packages into schedule activities where the basis for estimating, scheduling, executing, and monitoring and controlling the work of the project is easily supported and accomplished. This process documents the specific activities needed to fulfill the deliverables detailed in the WBS. By the end of this process, the project team will have defined a list of activities, their characteristics, and a milestones list.

Figure 3.12 shows the inputs, tools and techniques, and outputs of the Define Activities process.

**FIGURE 3.12** Define Activities process

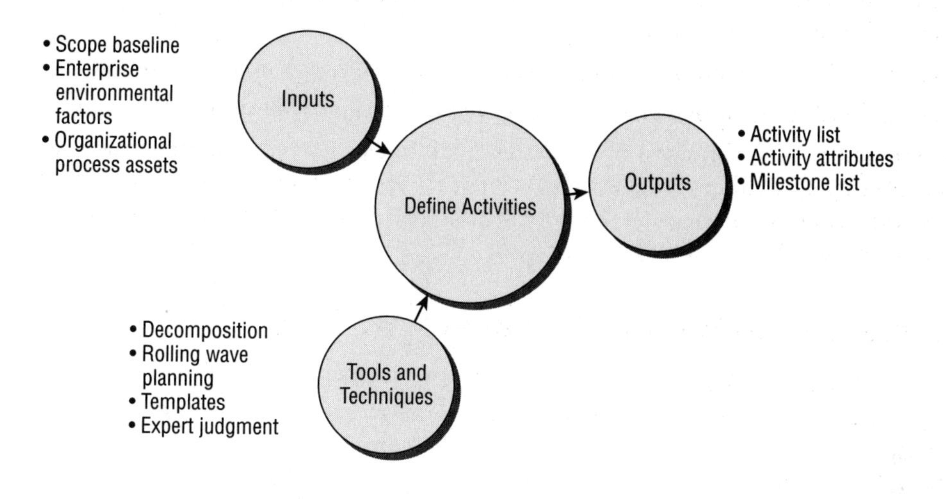

For more detailed information on the Define Activities process, see Chapter 4, "Creating the Project Schedule," of *PMP: Project Management Professional Exam Study Guide, 6th Edition*.

## Inputs of Define Activities

You should know the following inputs of the Define Activities process:

**Scope Baseline** Utilized and considered from within the scope baseline are the following:

- Work packages, from within the WBS
- Constraints and assumptions, from within the project scope statement

**Enterprise Environmental Factors** Utilized from within the enterprise environmental factors is the project management information system.

**Organizational Process Assets** The organizational process assets provide existing guidelines and policies, historical project documents (such as partial activity lists from previous projects), and a lessons-learned knowledge base, which are used for defining the project activities.

## Tools and Techniques of Define Activities

There are four tools and techniques within the Define Activities process that you can utilize:

- Decomposition
- Rolling wave planning

- Templates
- Expert judgment

**Decomposition** Decomposition in this process involves breaking the work packages into smaller, more manageable units of work called activities. *Activities* are individual units of work that must be completed to fulfill the deliverables listed in the WBS.

**Rolling Wave Planning** Rolling wave planning is a form of progressive elaboration that involves planning near-term work in more detail than future-term work.

**Templates** Templates utilized from within this process include activity lists from previous projects.

**Expert Judgment** Expert judgment helps define activities and involves project team members with prior experience developing project scope statements and WBSs.

## Outputs of Define Activities

The following are outputs of the Define Activities process:

**Activity List** Activity lists contain all the schedule activities that will be performed for the project, with a scope-of-work description of each activity and an identifier so that team members understand what the work is and how it is to be completed.

The schedule activities are individual elements of the project schedule.

**Activity Attributes** Activity attributes describe the characteristics of the activities and are an extension of the activity list. Activity attributes will change over the life of the project, as more information is known.

During the early stages of the project, activity attributes typically consist of the following:

- Activity ID
- WBS identification code it's associated with
- Activity name

As the project progresses, the following activity attributes may be added:

- Predecessor and successor activities
- Logical relationships
- Leads and lags
- Resource requirements
- Constraints and assumptions associated with the activity

**Milestone List** Milestones are major accomplishments of the project and mark the completion of major deliverables or some other key event in the project. The milestone list documents several things:

- Records the accomplishments
- Documents whether a milestone is mandatory or optional
- Becomes part of the project management plan
- Helps develop the project schedule

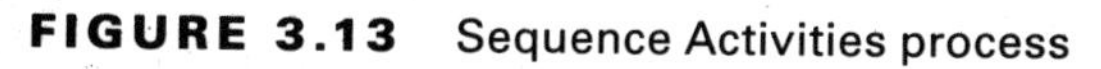

In practice, the Define Activities process and Sequence Activities process may be combined into one process or step.

## Sequence Activities

The Sequence Activities process takes the identified schedule activities from the Define Activities process, sequences them in logical order, and identifies any dependencies that exist among the activities. The interactivity of logical relationships must be sequenced correctly in order to facilitate the development of a realistic, achievable project schedule. The end of this process results in the creation of project schedule network diagrams, which visually show the sequence of activities and their dependencies.

Figure 3.13 shows the inputs, tools and techniques, and outputs of the Sequence Activities process.

**FIGURE 3.13** Sequence Activities process

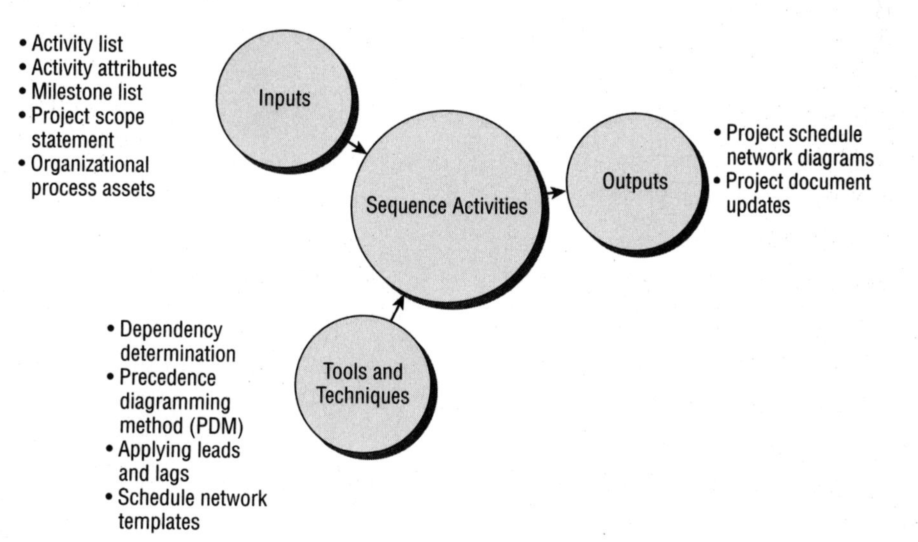

For more detailed information on the Sequence Activities process, see Chapter 4 of *PMP: Project Management Professional Exam Study Guide, 6th Edition.*

## Inputs of Sequence Activities

Within the Sequence Activities process, you will utilize the following inputs:

- Activity list
- Activity attributes
- Milestone list
- Project scope statement
- Organizational process assets

**Activity List** To sequence activities, you would first need to acquire the activity list as an input. The activity list includes the activities along with their identifiers and a description of the work's scope.

**Activity Attributes** Activity attributes may reveal information on the sequencing of activities, such as through predecessor or successor relationships.

**Milestone List** Milestone lists may contain milestones with scheduled dates. These dates might impact the sequencing of activities.

**Project Scope Statement** The product scope description within the project scope statement includes product details that may be used for sequencing activities.

**Organizational Process Assets** The project files found within the corporate knowledge base are utilized from within the organizational process assets for scheduling purposes.

## Tools and Techniques of Sequence Activities

You will need to be familiar with the following tools and techniques of the Sequence Activities process:

- Dependency determination
- Precedence diagramming method (PDM)
- Applying leads and lags
- Schedule network templates

**Dependency Determination** Dependencies are relationships between the activities in which one activity is dependent on another to complete an action, or perhaps an activity is dependent on another to start an action before it can proceed. Dependency determination is a matter of determining where those dependencies exist.

The following are three types of dependencies, defined by characteristics:

**Mandatory Dependencies** Mandatory dependencies, also known as hard logic or hard dependencies, are defined by the type of work being performed. The nature of the work itself dictates the order in which the activities should be performed.

**Discretionary Dependencies** Discretionary dependencies are defined by the project team. Discretionary dependencies are also known as preferred logic, soft logic, or preferential logic. These are usually process- or procedure-driven or best-practice techniques based on past experience.

**External Dependencies** External dependencies are external to the project. The *PMBOK® Guide* points out that even though the dependency is external to the project (and therefore a non-project activity), it impacts project activities.

**Precedence Diagramming Method** The precedence diagramming method (PDM) uses boxes or rectangles to represent the activities (called nodes). The nodes are connected with arrows showing the dependencies between the activities. This method is also called activity on node (AON). PDM uses only onetime estimates to determine duration.

The following information is displayed on a node:

- Activity name (required)
- Activity number (optional)
- Start and stop dates (optional)
- Due dates (optional)
- Slack time (optional)
- Any additional or relevant information (optional)

The following illustration shows a general example of a PDM, or AON.

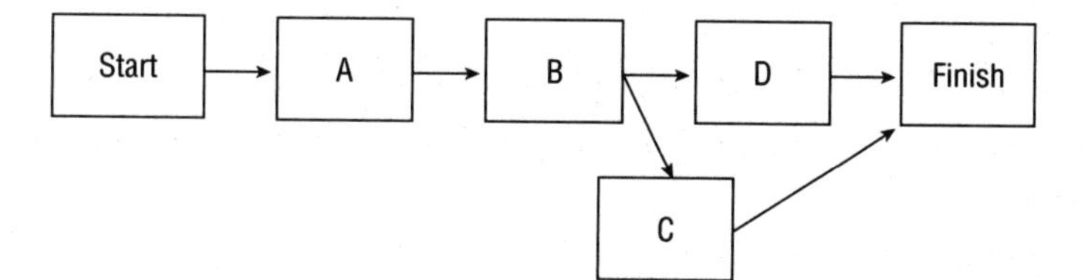


The PDM is further defined by the following four types of dependencies, also known as logical relationships:

**Finish-to-Start** The finish-to-start (FS) relationship is the most frequently used relationship. In this relationship, the predecessor—or *from* activity—must finish before the successor—or *to* activity—can start.

**Start-to-Finish** In the start-to-finish (SF) relationship, the predecessor activity must start before the successor activity can finish. This logical relationship is seldom used.

**Finish-to-Finish** In the finish-to-finish (FF) relationship, the predecessor activity must finish before the successor activity finishes.

**Start-to-Start** In the start-to-start (SS) relationship, the predecessor activity must start before the successive activity can start.

**Arrow Diagramming Method** The arrow diagramming method (ADM) technique isn't used nearly as often as PDM. ADM is visually the opposite of the PDM. The arrow diagramming method places activities on the arrows, which are connected to dependent activities with nodes. This method is also called activity on arrow (AOA). Characteristics of the ADM technique are as follows:

- ADM allows for more than onetime estimates to determine duration and uses only the finish-to-start dependency.
- Dummy activities may be plugged into the diagram to accurately display the dependencies.

The following illustration shows a general example of the ADM method.

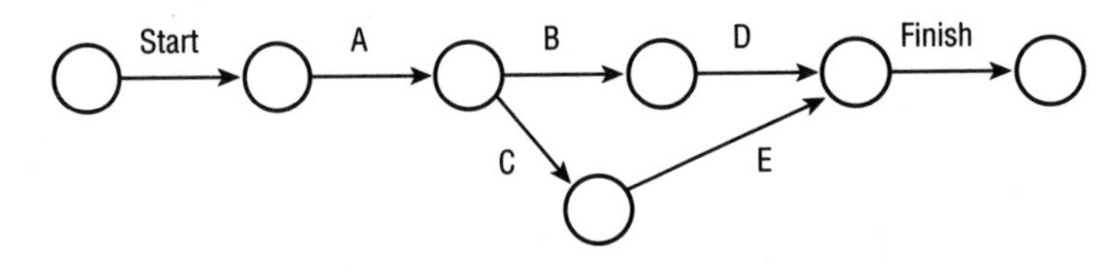


The following illustration is meant to help you remember the difference between PDM and ADM for the exam.

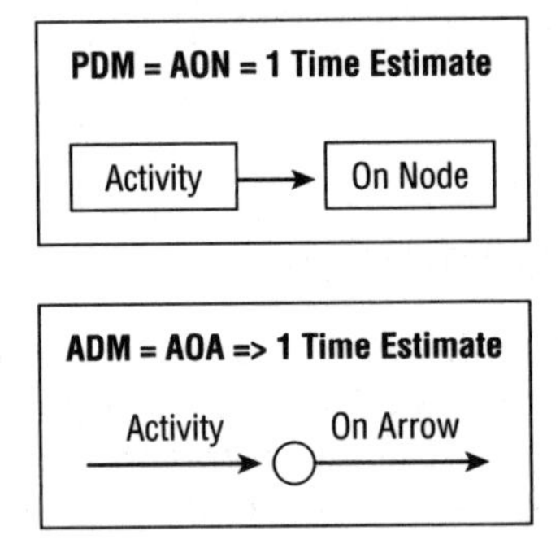


**Applying Leads and Lags** Leads and lags should be considered when determining dependencies:

**Leads** Leads speed up the successor activities and require time to be subtracted from the start date or the finish date of the activity you're scheduling.

**Lags** Lags delay successor activities (those that follow a predecessor activity) and require time added either to the start date or to the finish date of the activity being scheduled.

**Schedule Network Templates** Schedule network diagrams from previous similar projects can be used as templates.

### Outputs of Sequence Activities

There are two outputs of the sequence activities process that you should be familiar with:

**Project Schedule Network Diagrams** Like the WBS, the project schedule network diagrams might contain all the project details or contain only summary-level details, depending on the complexity of the project. Summary-level activities are a collection of related activities also known as hammocks. *Hammocks* are a group of related activities rolled up into a summary heading that describes the activities likely to be contained in that grouping.

**Project Document Updates** The following project documents may require updates as a result of this process:

- Activity lists
- Activity attributes
- Risk register (for an introduction to the risk register, see "Outputs of Identify Risks" later in this chapter)

Figure 3.14 shows the position of the Sequence Activities process in relation to the other time management processes.

**FIGURE 3.14** Order of Sequence Activities process

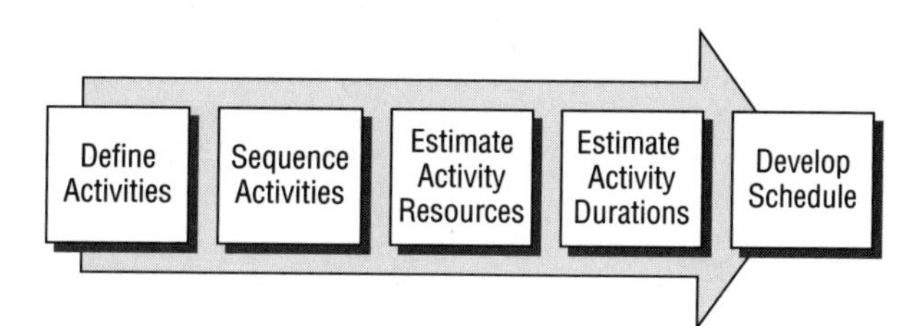


## Estimate Activity Resources

After the activities are sequenced, the next steps involve estimating the resources and estimating the durations of the activities so that they can be plugged into the project schedule. The Estimate Activity Resources process is concerned with determining the types and quantities of resources (both human and materials) needed for each schedule activity within a work package.

The term *resource* refers to all the physical resources required to complete the project. The *PMBOK® Guide* defines resources as people, equipment, and materials.

Figure 3.15 shows the inputs, tools and techniques, and outputs of the Estimate Activity Resources process.

**FIGURE 3.15** Estimate Activity Resources process

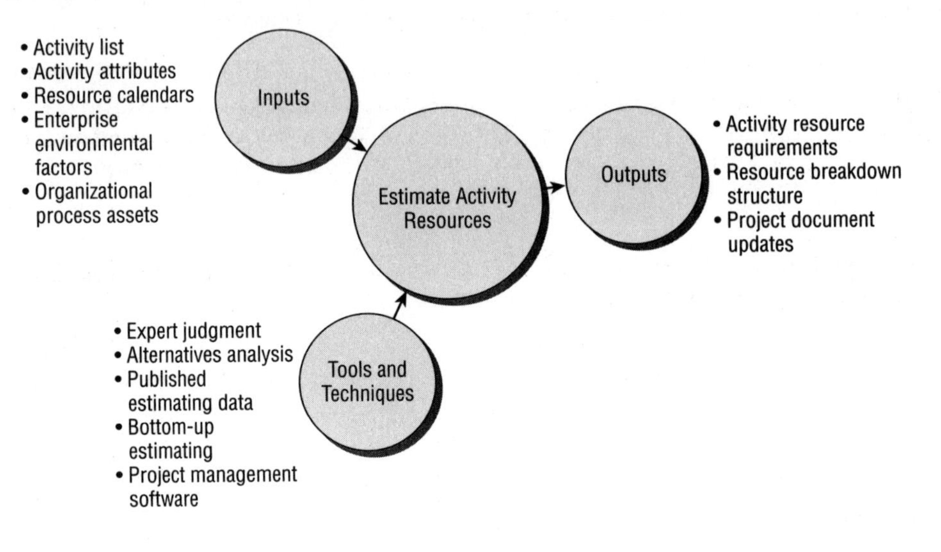

For more detailed information on the Estimate Activity Resources process, see Chapter 4 of *PMP: Project Management Professional Exam Study Guide, 6th Edition.*

## Inputs of Estimate Activity Resources

You should know the following inputs for the Estimate Activity Resources process:

- Activity list
- Activity attributes
- Resource calendars
- Enterprise environmental factors
- Organizational process assets

**Activity List** The activity list is necessary to know which activities will need resources.

**Activity Attributes** The activity attributes provide the details necessary to come up with an estimate of the resources needed per activity.

**Resource Calendars** Resource calendars describe the time frames in which resources are available and include the following:

- Skills
- Abilities
- Quantity
- Availability

Resource calendars also examine the quantity, capability, and availability of equipment and material resources that have a potential to impact the project schedule.

Resource calendars are an output of the Acquire Project Team and Conduct Procurements processes. Both of these processes are performed during the Executing process group. For additional information on these processes, see Chapter 4 within this book.

**Enterprise Environmental Factors** Information on the resource availability and skills can be obtained through enterprise environmental factors.

**Organizational Process Assets** Information from previous similar projects, and policies and procedures relating to staffing and equipment purchase and rental, are utilized from within the organizational process assets.

## Tools and Techniques of Estimate Activity Resources

There are several tools and techniques within the Estimate Activity Resources process:

- Expert judgment
- Alternatives analysis
- Published estimating data
- Bottom-up estimating
- Project management software

**Expert Judgment** Individuals with experience and knowledge of resource planning and estimating can be tapped for information and guidance utilized within this process.

**Alternatives Analysis** Alternatives analysis helps to make decisions about the possible resource types (such as expert or novice) and methods that are available to accomplish the activities. Make-or-buy analysis can also be used for decisions regarding resources.

**Published Estimating Data** Estimating data might include organizational guidelines, industry rates or estimates, production rates, and so on.

**Bottom-Up Estimating** Bottom-up estimating is used when an activity cannot be confidently estimated. With the help of experts, the activity is broken down into smaller components of work for estimating purposes and then rolled back up to the original activity level. This is an accurate means of estimating, but it can be time-consuming and costly.

**Project Management Software** Within this process, project management software can help in the following ways:

- Plan, organize, and estimate resource needs
- Document resource availability
- Create resource breakdown structures

- Create resource rates
- Create resource calendars

## Outputs of Estimate Activity Resources

Three outputs result from carrying out the Estimate Activity Resources process:

**Activity Resource Requirements** Activity resource requirements describe the types of resources and the quantity needed for each activity associated with a work package. The description should include the following:

- How estimates were determined
- Information used to form the estimate
- Assumptions made about the resources and their availability

Work package estimates are derived by taking a cumulative total of all the schedule activities within the work package.

**Resource Breakdown Structure** The resource breakdown structure (RBS) is a hierarchical structure that lists resources by category and type. Figure 3.16 illustrates a basic example of what an RBS may look like.

**FIGURE 3.16** Resource breakdown structure

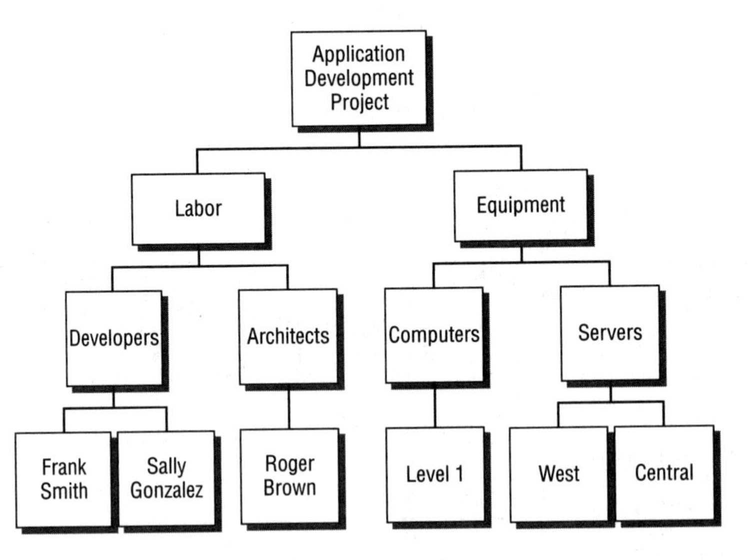


Here are some examples of categories:

- Labor
- Hardware

- Equipment
- Supplies

Here are some examples of types:

- Skill levels
- Quality grades
- Cost

**Project Document Updates** Project document updates include updates to the following items:

- Activity list
- Activity attributes
- Resource calendars

## Estimate Activity Durations

The Estimate Activity Durations process attempts to estimate the work effort, resources, and number of work periods needed to complete each activity. These are quantifiable estimates expressed as the number of work periods needed to complete a schedule activity. Estimates are progressively elaborated, typically starting at a fairly high level, and as more and more details are known about the deliverables and their associated activities, the estimates become more accurate.

Figure 3.17 shows the inputs, tools and techniques, and outputs of the Estimate Activity Durations process.

**FIGURE 3.17** Estimate Activity Durations process

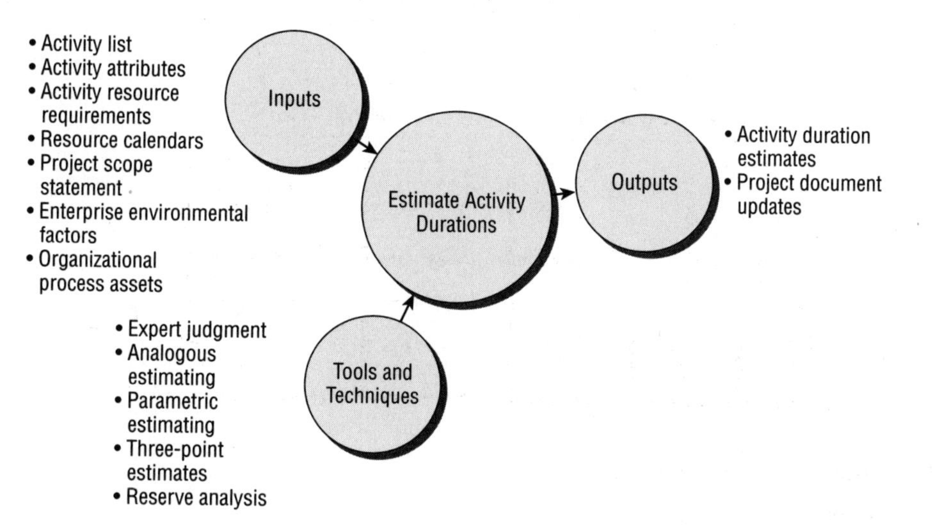

For more detailed information on the Estimate Activity Durations process, see Chapter 4 of *PMP: Project Management Professional Exam Study Guide, 6th Edition.*

## Inputs of Estimate Activity Durations

There are several inputs of the Estimate Activity Durations process that you should be familiar with:

- Activity list
- Activity attributes
- Activity resource requirements
- Resource calendars
- Project scope statement
- Enterprise environmental factors
- Organizational process assets

**Activity List** The list of activities is necessary to estimate the activity durations.

**Activity Attributes** Information included within the activity attributes will influence the activity duration estimates.

**Activity Resource Requirements** The availability of the assigned resources will impact the duration of the activities.

**Resource Calendars** Information utilized from the resource calendars that influences this process is as follows:

- Type, availability, and capability of human resources
- Type, availability, capability, and quantity of equipment and material resources

**Project Scope Statement** Constraints and assumptions are considered when estimating activity durations.

**Enterprise Environmental Factors** Enterprise environmental factors utilized include the following items:

- Internal reference data for estimating durations
- External reference data available commercially
- Defined productivity metrics

**Organizational Process Assets** The following organizational process assets are utilized:

- Historical information from previous similar projects, including duration information, project calendars, and lessons learned
- Scheduling methodology

## Tools and Techniques of Estimate Activity Durations

The five tools and techniques of the Estimate Activity Durations process are as follows:

- Expert judgment
- Analogous estimating
- Parametric estimating
- Three-point estimates
- Reserve analysis

**Expert Judgment** Expert judgment used includes staff members who will perform the activities and is based on their experience with past similar activities. When estimating durations, experts should consider the following:

- Resource levels
- Resource productivity
- Resource capability
- Risks
- Any other factors that can impact estimates

**Analogous Estimating** Analogous estimating is commonly used when little detail is available on the project. For a complete description of analogous estimating, see "Tools and Techniques of Estimate Costs" earlier in this chapter.

**Parametric Estimating** Parametric estimating is a quantitatively based estimating method that multiplies the quantity of work by the rate. It is considered to be a quick and low-cost estimating technique, with a good level of accuracy when used with actual historical data and current market conditions that take inflation or other factors into consideration. The best way to describe it is with an example.

Activity: install 15 10×10 drapes.

Average time to install 1 10×10 drape, based on previous experience: 30 minutes.

Estimate: Therefore, installing 15 drapes at an average 30-minute installation time per drape results in an estimated duration of 7.5 hours.

The *PMBOK® Guide* states that parametric estimating can also be used to determine time estimates for the entire project or portions of the project.

**Three-Point Estimates** Three-point estimates use the average of the following three estimates to result in a final estimate:

**Most Likely (ML)** The estimate assumes there are no disasters and the activity can be completed as planned.

**Optimistic (O)** This represents the fastest time frame in which your resource can complete the activity.

**Pessimistic (P)** The estimate assumes the worst happens and it takes much longer than planned to get the activity completed.

The concept of the three-point estimate comes from the Program Evaluation and Review Technique (PERT), which uses the following formula to determine the weighted average using the three duration estimates:

*(Optimistic* + 4 × (*Most Likely*) + *Pessimistic)* ÷ 6

In the following example, the Most Likely (ML) estimate = 15, the Optimistic (O) estimate = 11, and the Pessimistic (P) estimate = 19.

(11 + 4(15) + 19) ÷ 6

Thus, the activity duration = 15.

**Reserve Analysis** Reserve time—also called *buffer/time reserves* or *contingency reserve* in the *PMBOK® Guide*—is the portion of time added to the activity to account for schedule risk or uncertainty. To make sure the project schedule is not impacted, a reserve time of the original estimate is built in to account for the problems that may be encountered.

Original activity duration to install 15 10×10 drapes: 7.5 hours.

10 percent reserve time added: 45 minutes.

Estimate: Therefore, the new activity duration is adjusted to include the reserve, resulting in an estimated duration of 8.25 hours.

## Outputs of Estimate Activity Durations

There are two outputs of the Estimate Activity Durations process you should know:

- Activity duration estimates
- Project document updates

**Activity Duration Estimates** Activity duration estimates are estimates of the required work periods needed to complete the activity. This is a quantitative measure usually expressed in hours, weeks, days, or months.

Final estimates should contain a range of possible results, such as *X* hours ± 10 hours, or a percentage may be used to express the range.

Original activity duration to install 15 10×10 drapes: 7.5 hours.

Final estimate: 7.5 hours ± 1 hour. Therefore, the activity may take as little as 6.5 hours or as much as 8.5 hours.

**Project Document Updates** The following information may need to be revisited and updated as a result of this process:

- Activity attributes
- Assumptions made regarding resource availability and skill levels

## Develop Schedule

The purpose of the Develop Schedule process is to create the project schedule. Here, start and finish dates are created, activity sequences and durations are finalized, and the critical path is determined. The Develop Schedule process cannot be completed until the following processes of project planning have occurred:

- Collect requirements
- Define scope
- Create WBS
- Define activities
- Sequence activities
- Estimate activity durations
- Develop human resource plan

Figure 3.18 shows the inputs, tools and techniques, and outputs of the Develop Schedule process.

**FIGURE 3.18** Develop Schedule process

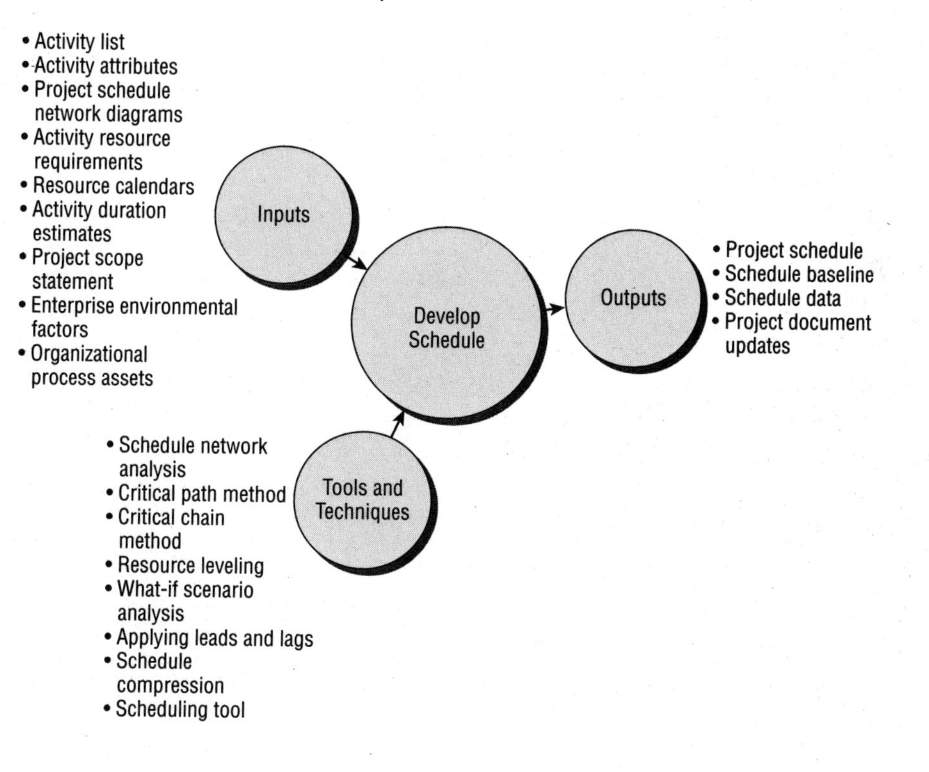

For more detailed information on the Develop Schedule process, see Chapter 4 of *PMP: Project Management Professional Exam Study Guide, 6th Edition.*

## Inputs of Develop Schedule

The Develop Schedule process has nine inputs you should be familiar with:

- Activity list
- Activity attributes
- Project schedule network diagrams
- Activity resource requirements
- Resource calendars
- Activity duration estimates
- Project scope statement
- Enterprise environmental factors
- Organizational process assets

**Activity List** Developing the project schedule requires the project activities, which are obtained from the activity list developed in the Define Activities process.

**Activity Attributes** Activity attributes provide the necessary details of the project activities, which will be utilized in creating the project schedule.

**Project Schedule Network Diagrams** The sequence of events will be important in creating the project schedule. The activity sequence and existing dependencies can be obtained from the project schedule network diagrams. For a more complete description of the project schedule network diagrams, see "Outputs of Sequence Activities" earlier in this chapter.

**Activity Resource Requirements** Activity resource requirements will provide the types and quantity of resources needed for creating the project schedule.

**Resource Calendars** Resource calendars display availability of project team members, which will help to avoid schedule conflicts.

**Activity Duration Estimates** Activity duration estimates will provide the time to complete each activity that is needed to assign start and finish dates.

**Project Scope Statement** Constraints and assumptions are utilized from within the project scope statement. The following time constraints are important to this process:

- Imposed dates to restrict the start or finish date of activities
- Key events/major milestones to ensure the completion of specific deliverables by a specific date

**Enterprise Environmental Factors** Existing holidays and other external information that could impact the project schedule may be utilized from within the enterprise environmental factors.

**Organizational Process Assets** Within the organizational process assets, project calendars may exist that provide information on working days and shifts that can be utilized in this process.

## Tools and Techniques of Develop Schedule

The Develop Schedule process has several tools and techniques you may use:

- Schedule network analysis
- Critical path method
- Critical chain method
- Resource leveling
- What-if scenario analysis
- Applying leads and lags
- Schedule compression
- Scheduling tool

**Schedule Network Analysis** Schedule network analysis involves calculating early and late start dates and early and late finish dates for project activities to generate the project schedule. In short, it generates the schedule through the use of the following analytical techniques and methods:

- Critical path method
- Critical chain method
- What-if scenario analysis
- Resource leveling

Calculations are performed without taking resource limitations into consideration, so the dates are theoretical. Resource limitations and other constraints are taken into consideration during the outputs of this process.

**Critical Path Method** Critical path method (CPM), a schedule network analysis technique, determines the amount of float, or schedule flexibility, for each of the network paths by calculating the earliest start date, earliest finish date, latest start date, and latest finish date for each activity. This technique relies on sequential networks and a single duration estimate for each activity. PDM can be used to perform CPM. Here is some information related to CPM that you should know:

**Critical Path** The critical path (CP) is generally the longest full path on the project. Any project activity with a float time that equals zero is considered a critical path task. The critical path can change under the following conditions:

- When activities become tasks on the critical path as a result of having used up all their float time
- When a milestone on the critical path is not met

**Float** There are two types of float time, also called slack time:

- Total float (TF), or the amount of time you can delay the earliest start of a task without delaying the ending of the project
- Free float (FF), or the amount of time you can delay the start of a task without delaying the earliest start of a successor task

**Calculating the Forward and Backward Pass** A forward pass is used to calculate the early start and early finish date of each activity on a network diagram. To calculate a forward pass, follow these steps:

1. Begin with the first activity.
2. The calculation of the first activity begins with an early start date of zero. Add the duration of the activity to determine the early finish.
3. The early start date of the next activity is the early finish date of the previous activity. Continue calculating the early start and early finish dates forward through all the network paths while following the existing dependencies.
4. When an activity has two connecting predecessors, the early start date would be the early finish of the predecessor that finishes last. (For example, suppose activity A has an early finish of 4 and activity B has an early finish of 5. If activities A and B both converge at activity C, then activity C would have an early start date of 5.)

A backward pass is used to calculate the late start and late finish date of each activity on a network diagram. To calculate a backward pass, follow these steps:

1. Begin with the last activity. The late finish will be the same as the early finish date.
2. Subtract the duration of the activity from the end date to calculate late start. This becomes the late finish of its predecessors.
3. Continue calculating the latest start and latest finish dates moving backward through all of the network paths.
4. When an activity has two connecting predecessors, the late finish date would be the late start of the activity that follows. (For example, if activity C has a late start of 7 and both activity A and activity B connect to activity C, the late finish of both activity A and B would be 7.)

Figure 3.19 summarizes a forward and backward pass.

**FIGURE 3.19** Forward and backward pass

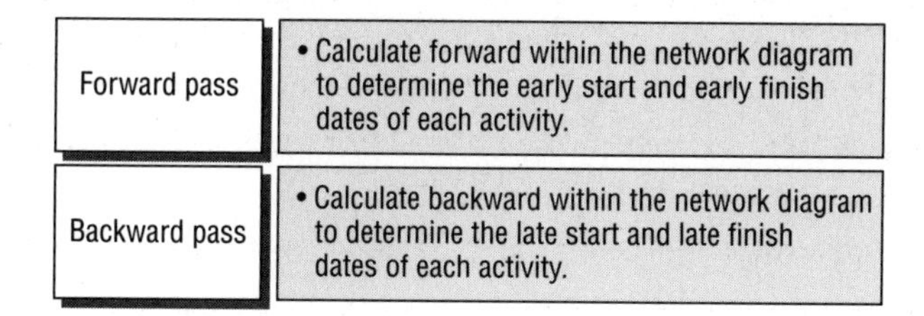


**Calculating the Critical Path** A critical path task is any task that cannot be changed without impacting the project end date. By definition, these are all tasks with zero float. To determine the CP duration of the project, add the duration of every activity with zero float.

Another way to determine the critical path is by determining the longest path within the network diagram. This can be done by adding the duration of all the activities within each path of the network.

Using Table 3.2, set up a network diagram using the activity number, activity description, dependency, and duration. Next, practice calculating the early start, early finish, late start, and late finish dates. Check your answers against the dates shown on the chart and by referencing Figure 3.20.

**FIGURE 3.20** Critical path diagram

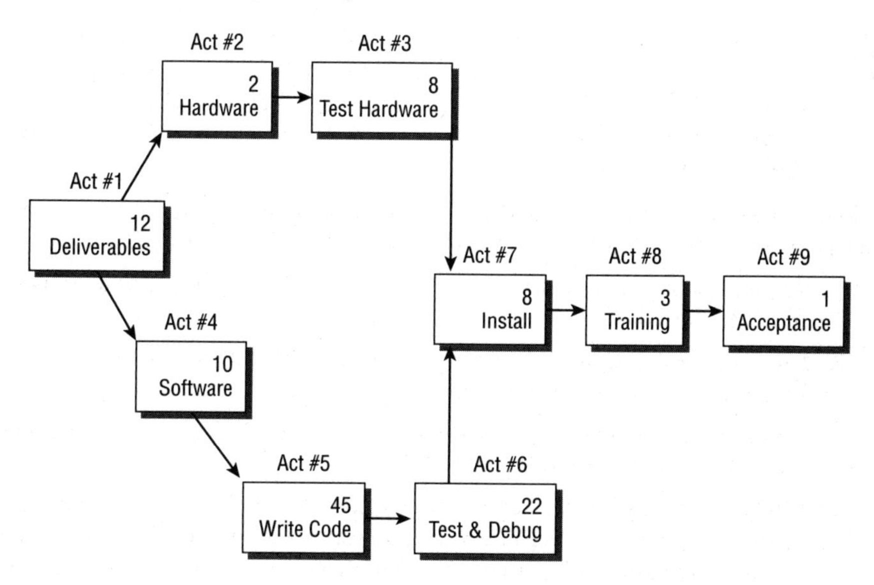

**TABLE 3.2** CPM calculation

| Activity Number | Activity Description | Dependency | Duration | Early Start | Early Finish | Late Start | Late Finish | Float/ Slack |
|---|---|---|---|---|---|---|---|---|
| 1 | Project Deliver-ables | — | 12 | 4/1 | 4/12 | 4/1 | 4/12 | 0 |
| 2 | Procure Hardware | 1 | 2 | 4/13 | 4/14 | 6/19 | 6/20 | 67 |
| 3 | Test Hardware | 2 | 8 | 4/15 | 4/22 | 6/21 | 6/28 | 67 |
| 4 | Procure Software Tools | 1 | 10 | 4/13 | 4/22 | 4/13 | 4/22 | 0 |
| 5 | Write Programs | 4 | 45 | 4/23 | 6/6 | 4/23 | 6/6 | 0 |
| 6 | Test and Debug | 5 | 22 | 6/7 | 6/28 | 6/7 | 6/28 | 0 |
| 7 | Install | 3, 6 | 8 | 6/29 | 7/6 | 6/29 | 7/6 | 0 |
| 8 | Training | 7 | 3 | 7/7 | 7/9 | 7/7 | 7/9 | 0 |
| 9 | Acceptance | 8 | 1 | 7/10 | 7/10 | 7/10 | 7/10 | 0 |

**Calculating Expected Value Using PERT** PERT and CPM are similar techniques: CPM uses the most likely duration to determine project duration, while PERT uses what's called expected value (or the weighted average). Expected value is calculated using the three-point estimates for activity duration.

For an explanation of how to calculate three-point estimates, see the tools and techniques of the Estimate Activity Durations process.

**Calculating Standard Deviation** The formula for standard deviation is as follows:

(Pessimistic – Optimistic) / 6

For data that fits a bell curve, the following is true:

- Work will finish within ± 3 standard deviations 99.73 percent of the time.
- Work will finish within ± 2 standard deviations 95.44 percent of the time.
- Work will finish within ± 1 standard deviation 68.26 percent of the time.

**Standard Deviation**

The higher the standard deviation is for an activity, the higher the risk. Since standard deviation measures the difference between the pessimistic and the optimistic times, a greater spread between the two, which results in a higher number, indicates a greater risk. Conversely, a low standard deviation means less risk.

One standard deviation gives you a 68 percent (rounded) probability, and two standard deviations give you a 95 percent (rounded) probability. For an example of calculating date ranges for project durations, see Chapter 4 of *PMP: Project Management Professional Exam Study Guide, 6th Edition.*

**Critical Chain Method** Critical chain method is a schedule network analysis technique that will modify the project schedule by accounting for limited or restricted resources. The modified schedule is calculated, and it often changes the critical path as a result of adding duration buffers, which are nonworking activities added to help manage the planned activity durations. The new critical path showing the resource restrictions is called the critical chain.

Critical chain uses both deterministic (step-by-step) and probabilistic approaches. The following are steps in the critical chain process:

1. Construct the schedule network diagram using activity duration estimates (you'll use nonconservative estimates in this method).
2. Define dependencies.
3. Define constraints.
4. Calculate critical path.
5. Enter resource availability into the schedule.
6. Recalculate for the critical chain.

A project buffer is placed at the end of the critical chain to help keep the finish date from slipping. After the buffers are added, the planned activities are then scheduled at their latest start and finish dates.

**Resource Leveling** Resource leveling, also called resource-based method, is used when resources are overallocated. It attempts to smooth out the peaks and valleys of total resource usage through strategic consumption of float while also addressing any overcommitment of individual resources through rescheduling or reassignment.

**Overallocated Resources** Resource leveling attempts to smooth out the resource assignments to get tasks completed without overloading the individual while trying to keep the project on schedule. This typically takes the form of allocating resources to critical path tasks first.

Here are some examples of resource leveling:

- Delaying the start of a task to match the availability of a key team member
- Adjusting the resource assignments so that more tasks are given to team members who are underallocated
- Requiring the resources to work mandatory overtime

**Reverse Resource Allocation Scheduling** Reverse resource allocation scheduling is a technique used when key resources are required at a specific point in the project and they are the only resources available to perform these activities. This technique requires the resources to be scheduled in reverse order to assign the key resources at the correct time.

Resource leveling can cause the original critical path to change.

**What-If Scenario Analysis** What-if scenario analysis uses different sets of activity assumptions to produce multiple project durations. Simulation techniques such as Monte Carlo analysis use a range of probable activity durations for each activity, and those ranges are then used to calculate a range of probable duration results for the project itself. Monte Carlo runs the possible activity durations and schedule projections many, many times to come up with the schedule projections and their probability, critical path duration estimates, and float time.

**Applying Leads and Lags** A lead accelerates the start date of an activity by the number of days specified, while a lag delays the start date of an activity. Leads and lags were first used in the Sequence Activities process, discussed earlier in this chapter.

**Schedule Compression** Schedule compression is a form of mathematical analysis that's used to shorten the project schedule without changing the project scope. To be effective, work compressed must be based on those activities that fall on the critical path. There are two types of schedule compression techniques:

**Crashing** Crashing is a compression technique that looks at cost and schedule tradeoffs. This involves adding resources to critical path tasks in order to shorten the length of the tasks and therefore the length of the project. Crashing the schedule can lead to increased risk, increased costs, and a change in the critical path.

**Fast Tracking** Fast tracking is performing two tasks in parallel that were previously scheduled to start sequentially. Fast tracking can increase project risk and cause the project team to have to rework tasks, and it only works for activities that can be overlapped.

**Scheduling Tool** The scheduling tools used are typically in the form of project management software programs. They will automate the mathematical calculations and perform resource-leveling functions.

## Outputs of Develop Schedule

Know the following outputs of the Develop Schedule process:

- Project schedule
- Schedule baseline
- Schedule data
- Project document updates

**Project Schedule** The project schedule details the start and finish dates for each project activity as well as the resource assignments. In *PMBOK® Guide* terms, the project schedule is considered preliminary until resources are assigned. The following are additional elements of the project schedule:

- The project schedule should be approved and signed off by stakeholders and functional managers.
- For functional organizations, confirmation that resources will be available as outlined in the schedule should be obtained.
- The schedule cannot be finalized until approval and commitment for the resource assignments outlined in it are received.
- Once approved, the schedule becomes the schedule baseline for the remainder of the project.

The following are various ways of displaying the project schedule:

**Project Schedule Network Diagrams** Project schedule network diagrams usually show the activity dependencies and critical path. They will work as schedule diagrams when the start and finish dates are added to each activity.

**Gantt Charts** Gantt charts are commonly used to display schedule activities. They may show activity sequences, activity start and end dates, resource assignments, activity dependencies, and the critical path. Figure 3.21 shows a simple example that plots various activities against time.

**FIGURE 3.21** Gantt chart

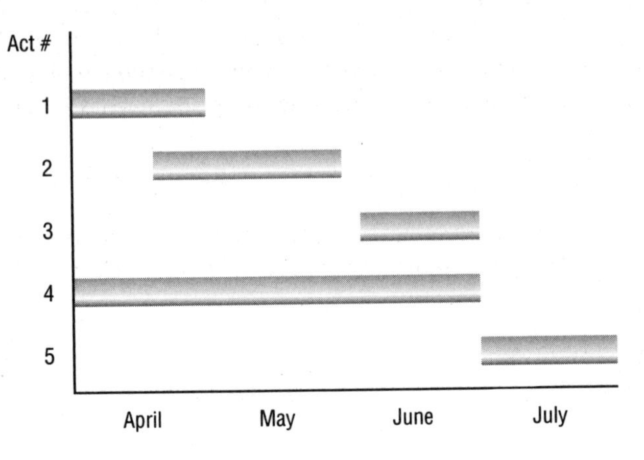


**Milestone Charts** Milestone charts mark the completion of major deliverables or some other key events in the project. They may be displayed in a bar chart form, similar to a Gantt chart, or use a simple table format, as shown in Table 3.3. As the milestones are met, the Actual Date column within the table is filled in.

**TABLE 3.3** Milestone chart

| Milestone | Scheduled Date | Actual Date |
|---|---|---|
| Sign-off on deliverables | 4/12 | 4/12 |
| Sign-off on hardware test | 4/22 | 4/25 |
| Programming completed | 6/06 | |
| Testing completed | 6/28 | |
| Acceptance and sign-off | 7/10 | |
| Project closeout | 7/10 | |

**Schedule Baseline** The schedule baseline can be described as the final, approved version of the project schedule with baseline start and baseline finish dates and resource assignments. The *PMBOK® Guide* notes that the schedule baseline is a designated version of the project schedule that's derived from the schedule network analysis. The approved project schedule becomes a part of the project management plan.

**Exam Note**

For the exam, remember that the project schedule is based on the timeline (derived from the activity), the scope document (to help keep track of major milestones and deliverables), and resource plans.

**Schedule Data** The schedule data refers to documenting the supporting data for the schedule. At least the following elements must be included within the schedule data:

- Milestones
- Schedule activities
- Activity attributes
- Assumptions
- Constraints
- Any other information that doesn't fit into the other categories
- Resource histograms (suggested by the *PMBOK® Guide*)

**Project Document Updates** Updates may be made to the following project documents:

- Activity resource requirements document
- Activity attributes
- Calendars
- Risk register

**Exam Essentials**

**Know the difference between the precedence diagramming method (PDM) and the arrow diagramming method (ADM).** PDM uses boxes or rectangles to represent the activities (called *nodes*) that are connected with arrows showing the dependencies between the activities. ADM places activities on the arrows (called *activity on arrow*), which are connected to dependent activities with nodes.

**Be able to describe the purpose of the Estimate Activity Resources process.** The purpose of Estimate Activity Resources is to determine the types and quantities of resources (human, equipment, and materials) needed for each schedule activity within a work package.

**Be able to name the tools and techniques of Estimate Activity Durations.** The tools and techniques of Estimate Activity Durations are expert judgment, analogous estimating, parametric estimating, and reserve analysis.

**Be able to define the difference between analogous estimating, parametric estimating, and bottom-up estimating.** Analogous estimating is a top-down technique that uses expert judgment and historical information. Parametric estimating multiplies the quantity of work by the rate. Bottom-up estimating performs estimates for each work item and rolls them up to a total.

**Be able to calculate the critical path.** The critical path includes the activities whose durations add up to the longest path of the project schedule network diagram. Critical path is calculated using the forward pass, backward pass, and float calculations.

**Be able to define a critical path task.** A critical path task is a project activity with zero float.

**Be able to describe and calculate PERT duration estimates.** This is a weighted average technique that uses three estimates: optimistic, pessimistic, and most likely. The formula is as follows:

$$(\text{Optimistic} + (4 \times \text{Most Likely}) + \text{Pessimistic}) \div 6$$

**Be able to name the schedule compression techniques.** The schedule compression techniques are crashing and fast tracking.

**Be able to describe a critical chain.** The critical chain is the new critical path in a modified schedule that accounts for limited resources.

**Be able to name the outputs of the Develop Schedule process.** The outputs are project schedule, schedule baseline, schedule data, and project document updates.

# Developing a Human Resource Plan

The project manager creates a human resource plan by carrying out a process called Develop Human Resource Plan. Through this process, roles and responsibilities of the project team members are defined and the strategies for utilizing and managing the project team are documented. The high-level purpose of the plan itself is to create an effective project organization structure.

The Develop Human Resource Plan process documents the roles and responsibilities of individuals or groups for various project elements and then documents the reporting relationships for each. Reporting relationships can be assigned to groups as well as to individuals, and the groups or individuals might be internal or external to the organization or a combination of both. This process will result in the creation of the human resource plan, which considers factors such as the availability of resources, skill levels, training needs, and more.

For more detailed information on the Develop Human Resource Plan process, see Chapter 7, "Planning Project Resources," of *PMP: Project Management Professional Exam Study Guide, 6th Edition*.

Figure 3.22 shows the inputs, tools and techniques, and outputs of the Develop Human Resource Plan process.

**FIGURE 3.22** Develop Human Resource Plan process

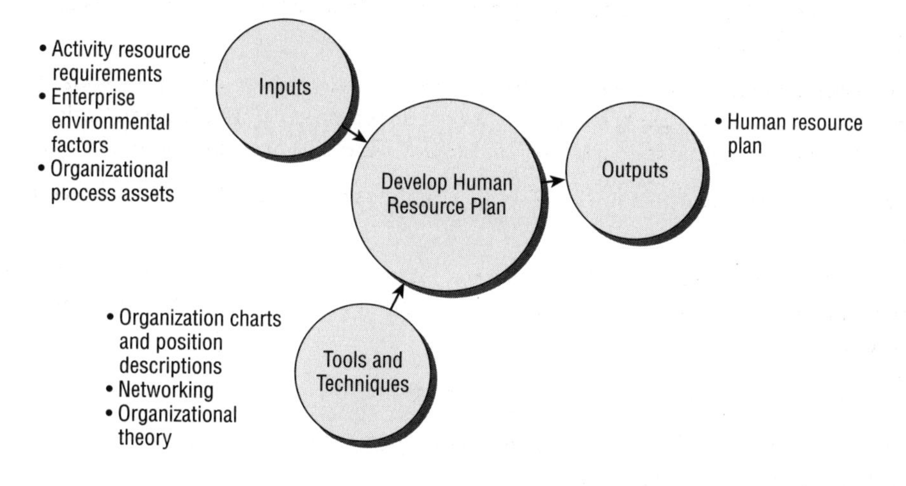


## Inputs of Develop Human Resource Plan

Inputs of the Develop Human Resource Plan process you should know are as follows:

**Activity Resource Requirements** Human resources are needed to perform and complete the activities outlined in the activity resource requirements output of the Estimate Activity Resources process. During the Develop Human Resource Plan process, these resources are defined in further detail.

**Enterprise Environmental Factors** Enterprise environmental factors play a key role in determining human resource roles and responsibilities. The following is a list of factors that are used:

- Organizational factors
- Existing human resources and marketplace conditions
- Personnel policies
- Technical factors
- Interpersonal factors
- Location and logistics
- Political factors
- Organizational structures
- Collective bargaining agreements
- Economic conditions

**Organizational Process Assets** The following three elements of the organizational process assets are used in this process:

- Organizational processes and standardized role descriptions
- Templates and checklists
- Historical information

# Tools and Techniques of Develop Human Resource Plan

The following are tools and techniques of the Develop Human Resource Plan process that you should be familiar with:

- Organization charts and position descriptions
- Networking
- Organizational theory

## Organization Charts and Position Descriptions

The following information is presented in organization charts and position descriptions:

**Hierarchical Charts** Hierarchical charts, like a WBS, are designed in a top-down format. Two types of hierarchical charts that can be used are OBS and RBS:

- Organization breakdown structure (OBS), which displays departments, work units, or teams in an organization and their respective work packages
- Resource breakdown structure (RBS), which breaks down the work of the project according to the types of resources needed

**Matrix-Based Charts** Matrix-based charts are used to show the type of resource and the responsibility that resource has in the project. There are a couple of types of matrix-based charts that can be used:

- Responsibility assignment matrix (RAM) is displayed as a chart with resource names listed in each column and project phases or WBS elements listed as the rows. Indicators in the intersections show where the resources are needed.
- A RACI chart is a type of RAM. Table 3.4 shows a sample portion of an RACI chart for a software development team. In this example, the RACI chart shows the roles and expectations of each participant.

The letters in the acronym RACI are the designations shown in Table 3.4:

R = Responsible for performing the work

A = Accountable, the one who is responsible for producing the deliverable or work package and approves or signs off on the work

C = Consult, someone who has input to the work or decisions

I = Inform, someone who must be informed of the decisions or results

**TABLE 3.4** Sample RAM

| | **Olga** | **Rae** | **Jantira** | **Nirmit** |
|---|---|---|---|---|
| **Design** | R | A | C | C |
| **Test** | I | R | C | A |
| **Implement** | C | I | R | A |

R = Responsible, A = Accountable, C = Consult, I = Inform

**Text-Oriented Formats** Text-oriented formats are used when there is a significant amount of detail to record. These are also called position descriptions or role-responsibility-authority forms. Text-oriented formats typically provide the following information:

- Role
- Responsibility
- Authority of the resource

**Other Sections of the Project Management Plan** Other subsidiary plans of the project management plan might also describe roles and responsibilities.

### Networking

Networking in this context means human resource networking. According to the *PMBOK® Guide*, several types of networking activities exist:

- Proactive communication
- Lunch meetings
- Informal conversations
- Trade conferences

### Organizational Theory

Organizational theory refers to all the theories that attempt to explain what makes people, teams, and work units perform.

## Outputs of Develop Human Resource Plan

There is only one output of the Develop Human Resource Plan process: the human resource plan.

The human resource plan documents how human resources should be defined, staffed, managed and controlled, and released from the project when their activities are complete. This plan is meant to clearly outline the roles and responsibilities of the project team and to create an effective project organization structure. As a result, guidance is provided on how resources are to be utilized and managed throughout the project.

This output has the following three components:

**Roles and Responsibilities** The roles and responsibilities component includes the list of roles and responsibilities for the project team. It can take the form of the RAM or RACI, or the roles and responsibilities can be recorded in text format. The following key elements should be included in the roles and responsibilities documentation:

- *Role* describes which part of the project the individuals or teams are accountable for. This should also include a description of authority levels, responsibilities, and what work is not included as part of the role.
- *Authority* describes the amount of authority the resource has to make decisions, dictate direction, and approve the work.
- *Responsibility* describes the work required to complete the project activities.
- *Competency* describes the skills and ability needed to perform the project activities.

**Project Organizational Charts** The project organizational charts display project team members and their reporting relationships in the project.

**Staffing Management Plan** The staffing management plan documents how and when human resources are introduced to the project and the criteria for releasing them. The staffing management plan should be updated throughout the project. The following elements should be considered and included in the staffing management plan:

- Staff acquisition describes how team members are acquired, where they're located, and the costs for specific skills and expertise.
- Resource calendars describe the time frames in which the resources will be needed on the project and when the recruitment process should begin.
- Staff release plan describes how project team members will be released at the end of their assignment, including reassignment procedures.
- Training needs describe any training plans needed for team members who don't have the required skills or abilities to perform project tasks.
- Recognition and rewards describe the systems used to reward and reinforce desired behavior.
- Compliance details regulations that must be met and any human resource policies the organization has in place that deal with compliance issues.
- Safety includes safety policies and procedures that are applicable to the project or industry and should be included in the staffing management plan.

### Exam Essentials

**Be able to define the purpose of the Develop Human Resource Plan process.** Develop Human Resource Plan involves determining roles and responsibilities, reporting relationships for the project, and creating the staffing management plan, which describes how team members are acquired and the criteria for their release.

# Developing a Communications Management Plan

Communications plays an important role in effectively and efficiently carrying out a project—with successful results! Like other plans, it should be as detailed as needed for the project at hand and should be based on the project organization structure and stakeholder requirements. This plan is responsible for managing the flow of project information and is created out of the Plan Communications process.

Keep in mind that, according to PMI, a good project manager spends up to 90 percent of their time communicating. Therefore, the communications management plan should be well thought out and clearly documented.

The Plan Communications process involves determining the communication needs of the stakeholders by defining the types of information needed, the format for communicating the information, how often it's distributed, and who prepares it. All of this is documented in the communications management plan, which is an output of this process. The total number of existing communication channels is also calculated in this process.

Figure 3.23 shows the inputs, tools and techniques, and outputs of the Plan Communications process.

**FIGURE 3.23** Plan Communications process

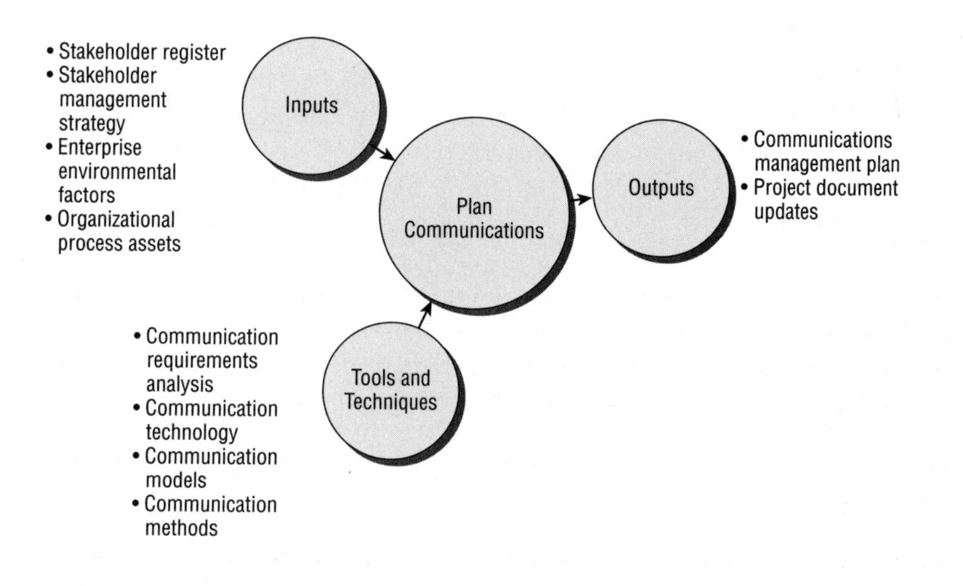

For more detailed information on the Plan Communications process, see Chapter 5 of *PMP: Project Management Professional Exam Study Guide, 6th Edition*.

## Inputs of Plan Communications

Know the following inputs of the Plan Communications process:

- Stakeholder register
- Stakeholder management strategy
- Enterprise environmental factors
- Organizational process assets

**Stakeholder Register** The stakeholder register is a list of all the project stakeholders and contains additional information, including their potential influence on the project and their classifications.

**Stakeholder Management Strategy** The stakeholder management strategy defines the level of participation needed for each stakeholder and potential strategies used to gain their support or reduce the negative impacts or obstacles they could present to the project.

**Enterprise Environmental Factors** All enterprise environmental factors can be utilized within the Plan Communications process. Special attention should be given to the organizational structure and culture as well as external stakeholder requirements. This information will be key to managing the flow of the project's information.

**Organizational Process Assets** The *PMBOK® Guide* notes that all the elements described in the enterprise environmental factors can be utilized within this process. Particular consideration should be given to lessons learned and historical information.

The *PMBOK® Guide* notes that there is a difference between effective and efficient communication. Effective communication refers to providing the information in the right format for the intended audience at the right time. Efficient communication refers to providing the appropriate information at the right time—that is, only the information that's needed at the time.

## Tools and Techniques of Plan Communications

The Plan Communications process includes the following tools and techniques:

- Communication requirements analysis
- Communication technology

- Communication models
- Communication methods

## Communication Requirements Analysis

Communication requirements analysis involves analyzing and determining the communication needs of the project stakeholders. According to the *PMBOK® Guide*, there are several sources of information you can examine to help determine these needs:

- Company and departmental organizational charts
- Stakeholder responsibility relationships
- Other departments and business units involved in the project
- The number of resources involved in the project and where they're located in relation to project activities
- Internal needs that the organization may need to know about the project
- External needs that organizations such as the media, government, or industry groups might have that require communication updates
- Stakeholder information (documented in the stakeholder register and stakeholder management strategy outputs of Identify Stakeholders)

The preceding list goes through analysis to make certain that information that is valuable to the stakeholders is being supplied. It is important to know the number of communication channels, also known as lines of communication, that exist. Figure 3.24 shows an example of a network model with six stakeholders included.

**FIGURE 3.24** Network communication model

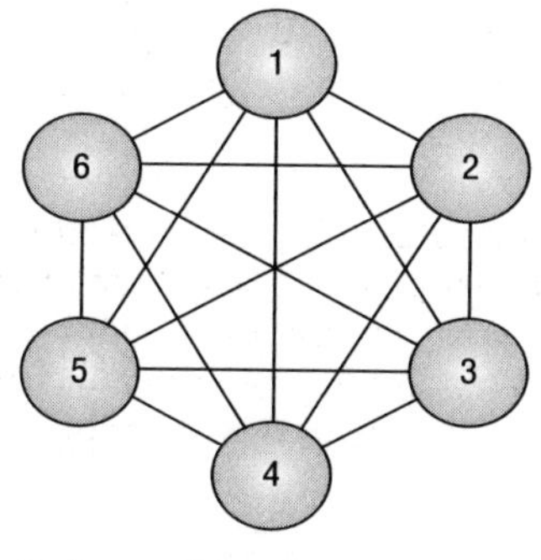

The nodes are the participants, and the lines show the connection between them all. The formula for calculating the lines of communication is as follows (where *n* represents total participants):

$$\text{lines of communication} = n(n - 1) \div 2$$

Figure 3.24 shows six participants. When you plug the total participants into the formula, the result is a total of 15 communication channels:

$$6(6 - 1) \div 2 = 15$$

## Communication Technology

Communication technology examines the methods (or technology) used to communicate the information to, from, and among the stakeholders. This tool and technique examines the technology elements that might affect project communications.

The following are examples of methods used when communicating:

- Written
- Spoken
- Email
- Formal status reports
- Meetings
- Online databases
- Online schedules

## Communication Models

Communication models depict how information is transmitted from the sender and how it's received by the receiver. According to the *PMBOK® Guide*, the key components of a communication model are as follows:

**Encode** Encoding the message means to put information or thoughts into a language that the receiver will understand.

**Message and Feedback Message** The result or output of the encoding.

**Medium** The method used to communicate, such as written, oral, or email.

**Noise** Anything that keeps the message from being either transmitted or understood.

**Decode** The receiver translates the information that was sent by the sender.

The sender is responsible for encoding the message, selecting the medium, and decoding the feedback message. The receiver is responsible for decoding the original message from the sender and encoding and sending the feedback message. Figure 3.25 shows this dynamic through the basic communication model.

**FIGURE 3.25** Communication model

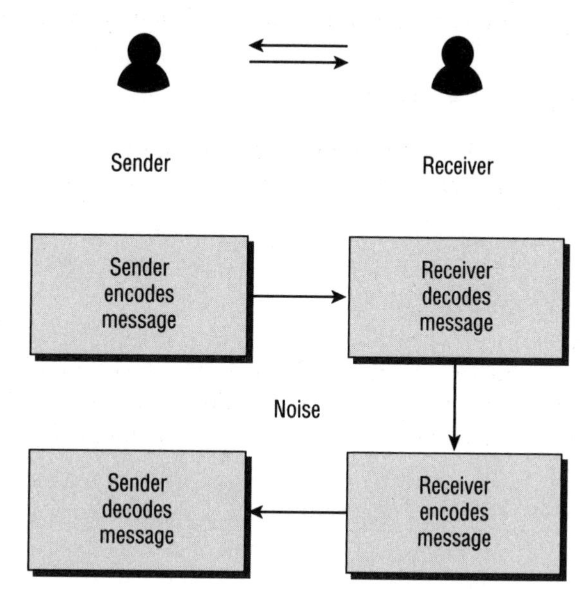


## Communication Methods

Communication methods refer to how the project information is shared among the stakeholders. According to the *PMBOK® Guide*, there are three classifications of communication methods:

**Interactive Communication** Interactive communication involves multidirectional communication where two or more parties must exchange thoughts or ideas. This method includes videoconferencing, phone or conference calls, and meetings.

**Push Communications** Push communications refers to sending information to intended receivers. It includes methods such as letters, memos, reports, emails, voicemails, and so on. This method assures that the communication was sent but is not concerned with whether it was actually received or understood by the intended receivers.

**Pull Communications** The likely recipients of the information access the information themselves using methods such as websites, e-learning sites, knowledge repositories, shared network drives, and so on.

# Outputs of Plan Communications

The following two outputs result from carrying out the Plan Communications process:

**Communications Management Plan** The communications management plan documents the following:

- Types of information needs the stakeholders have
- When the information should be distributed
- How the information will be delivered

According to the *PMBOK® Guide*, the communications management plan typically describes the following elements:

- The communication requirements of each stakeholder or stakeholder group
- Purpose for communication
- Frequency of communications, including time frames for distribution
- Name of the person responsible for communicating information
- Format of the communication and method of transmission
- Method for updating the communications management plan
- Glossary of common terms

The information that will be shared with stakeholders and the distribution methods are based on the following items:

- Needs of the stakeholders
- Project complexity
- Organizational policies

**Project Document Updates** The following updates may be required as a result of performing this process:

- Project schedule
- Stakeholder register
- Stakeholder management strategy

**Exam Essentials**

**Be able to describe the purpose of the communications management plan.** The communications management plan determines the communication needs of the stakeholders. It documents what information will be distributed, how it will be distributed, to whom it will be distributed, and the timing of the distribution.

# Developing a Procurement Management Plan

In many cases, some resources will need to be obtained externally (meaning outside of the project's organization). When this is the case, a procurement management plan will be needed. This plan is created out of the Plan Procurements process, which is also responsible for producing other key procurement documents.

The procurement management plan should be based on the project's scope and schedule and is responsible for guiding the procurement-related processes and ensuring that the required project resources will be available when and as needed.

Plan Procurements is a process of identifying what goods or services will be purchased from outside the organization. This process addresses the make-or-buy decision and when acquired goods or services will be needed. By the end of this process, a procurement management plan will have been created, make-or-buy decisions will have been made, and a contract statement of work will have been drafted.

Figure 3.26 shows the inputs, tools and techniques, and outputs of the Plan Procurements process.

**FIGURE 3.26** Plan Procurements process

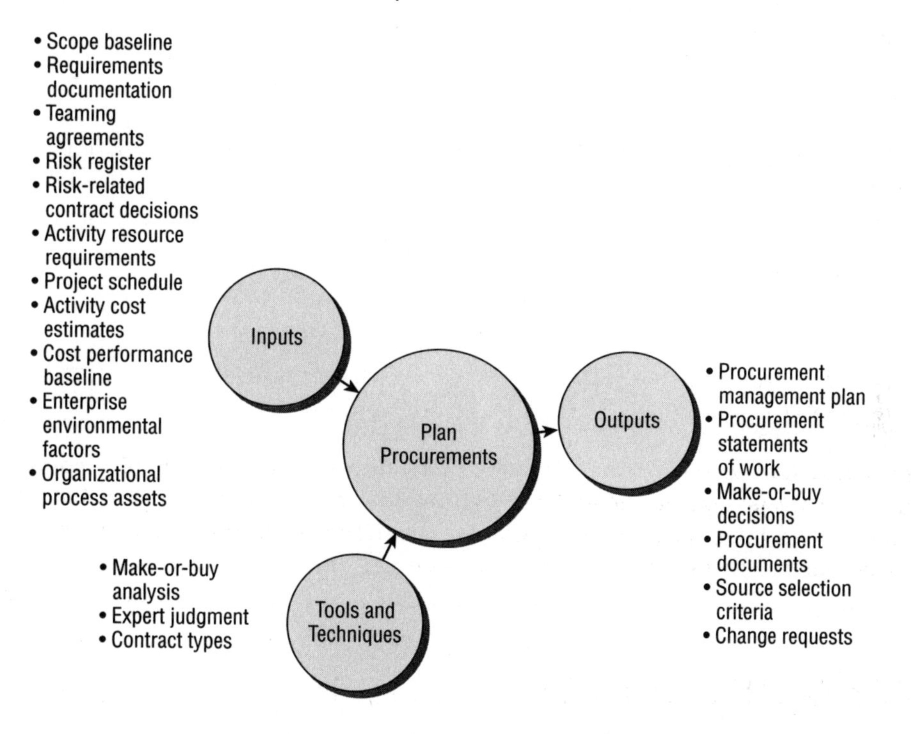

For more detailed information on the Plan Procurements process, see Chapter 7 of *PMP: Project Management Professional Exam Study Guide, 6th Edition*.

## Inputs of Plan Procurements

There are many inputs to the Plan Procurements process that you should be familiar with:

- Scope baseline
- Requirements documentation
- Teaming agreements
- Risk register
- Risk-related contract decisions
- Activity resource requirements
- Project schedule
- Activity cost estimates
- Cost performance baseline
- Enterprise environmental factors
- Organizational process assets

**Scope Baseline** The Plan Procurements process utilizes the following items from within the project scope statement, which is included within the scope baseline:

- Description of the need for the project
- Lists of deliverables
- Acceptance criteria
- Constraints
- Assumptions
- Product scope description (if applicable)

As part of the scope baseline, the WBS and WBS dictionary identify the deliverables and describe the work required for each element of the WBS.

**Requirements Documentation** The requirements documentation documents details of the project requirements. For a more detailed description, see "Outputs of Collect Requirements" earlier in this chapter.

**Teaming Agreements** Teaming agreements are contractual agreements between multiple parties that are forming a partnership or joint venture to work on the project. They are often used when two or more vendors form a partnership to work together on a particular project. If teaming agreements are used, the following should be predefined:

- Scope of work
- Requirements for competition
- Buyer and seller roles
- Any other important project concerns identified

**Risk Register** The risk register guides the project team in determining the types of services or goods needed for risk management.

**Risk-Related Contract Decisions** Risk-related contract decisions determine which activities will be addressed through the acquirement of insurance, services, and so forth to shift party responsibility as necessary.

**Activity Resource Requirements** The availability of the assigned resources will impact the procurement process and will aid in the make-or-buy decision.

**Project Schedule** The project schedule is necessary for determining the start and finish dates of activities that will be completed through a third party.

**Activity Cost Estimates** Activity cost estimates are necessary for determining the budget allotted for completing project activities, which will influence make-or-buy decisions and services or goods acquired.

**Cost Performance Baseline** The cost performance baseline shows the planned budget over the life span of the project.

**Enterprise Environmental Factors** Marketplace conditions are the key element of enterprise environmental factors used within this process.

**Organizational Process Assets** The organization's guidelines, policies, and procedures (including any procurement policies and guidelines) are the elements of the organizational process assets used in this process.

The project manager and the project team will be responsible for coordinating all the organizational interfaces for the project, including technical, human resource, purchasing, and finance.

## Tools and Techniques of Plan Procurements

There are three tools and techniques of the Plan Procurements process that you should know:

- Make-or-buy analysis
- Expert judgment
- Contract types

**Make-or-Buy Analysis** Make-or-buy analysis is concerned with whether it is more cost-effective to buy or lease the products and services or more cost-effective for the organization to produce the goods and services needed for the project. Costs should include both direct costs and indirect costs.

Other considerations in make-or-buy analysis are as follows:

- Capacity issues
- Skills
- Availability
- Trade secrets

Make-or-buy analysis is considered a general management technique and concludes with the decision to do one or the other.

**Expert Judgment** Expert judgment can be utilized within the Plan Procurements process.

**Contract Types** A contract is a compulsory agreement between two or more parties and is used to acquire products or services from outside the organization. Contracts are enforceable by law and require an offer and an acceptance, with money typically exchanged for goods or services. As reflected in Figure 3.27, there are different types of contracts for different purposes.

**FIGURE 3.27** Contract types

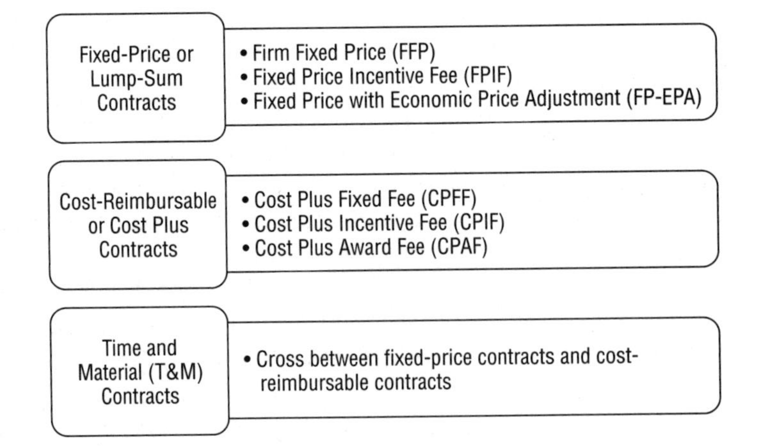


The *PMBOK® Guide* divides contracts into three categories:

**Fixed-Price, or Lump-Sum, Contracts** Fixed-price contracts (also referred to as lump-sum contracts) can either set a specific, firm price for the goods or services rendered (known as a firm fixed price contract) or can include incentives for meeting or exceeding certain contract deliverables. There are three types of fixed-price contracts:

- In a firm fixed price (FFP) contract, the buyer and seller agree on a well-defined deliverable for a set price. In this kind of contract, the biggest risk is borne by the seller. The seller assumes the risks of increasing costs, nonperformance, or other problems.
- Fixed price incentive fee (FPIF) contracts are another type of fixed-price contract. The difference is that the contract includes an incentive—or bonus—for early completion or for some other agreed-upon performance criterion that meets or exceeds contract specifications. Some of the risk is borne by the buyer as opposed to the firm fixed price contract, where most of the risk is borne by the seller.

- A fixed price with economic price adjustment (FP-EPA) contract allows for adjustments due to changes in economic conditions, like cost increases or decreases, inflation, and so on. These contracts are typically used when the project spans many years. This type of contract protects both the buyer and seller from economic conditions that are outside of their control.

**Cost-Reimbursable, or Cost Plus, Contracts** With cost-reimbursable contracts, the allowable costs associated with producing the goods or services are charged to the buyer. All the costs the seller takes on during the project are charged back to the buyer; thus, the seller is reimbursed. Cost-reimbursable contracts carry the highest risk for the buyer because the total costs are uncertain and are used most often when the project scope contains a lot of uncertainty or for projects that have large investments early in the project life. The following list includes four types of cost-reimbursable contracts:

- Cost plus fixed fee (CPFF) contracts charge back all allowable project costs to the seller and include a fixed fee upon completion of the contract. The fee is always firm in this kind of contract, but the costs are variable. The seller doesn't necessarily have a lot of motivation to control costs with this type of contract.
- Cost plus incentive fee (CPIF) is the type of contract in which the buyer reimburses the seller for the seller's allowable costs and includes an incentive for meeting or exceeding the performance criteria laid out in the contract. A possibility of shared savings between the seller and buyer exists if performance criteria are exceeded.
- Cost plus award fee (CPAF) is the riskiest of the cost plus contracts for the seller. In a CPAF contract, the seller will recoup all the costs expended during the project, but the award fee portion is subject to the sole discretion of the buyer.

**Time and Material (T&M) Contracts** T&M contracts are a cross between fixed-price and cost-reimbursable contracts. The full amount of the material costs is not known at the time the contract is awarded. T&M contracts are most often used when human resources with specific skills are needed and when the scope of work needed for the project can be quickly defined. Time and material contracts include the following characteristics:

- This type of contract resembles a cost-reimbursable contract because the costs will continue to grow during the contract's life and are reimbursable to the contractor. The buyer bears the biggest risk in this type of contract.
- This type of contract resembles fixed-price contracts when unit rates are used. Unit rates might be used to preset the rates of certain elements or portions of the project.

## Outputs of Plan Procurements

Know the following outputs of the Plan Procurements process:

- Procurement management plan
- Procurement statements of work

- Make-or-buy decisions
- Procurement documents
- Source selection criteria
- Change requests

**Procurement Management Plan** The procurement management plan details how the procurement process will be managed. According to the *PMBOK® Guide*, it includes the following information:

- Types of contracts to use
- Authority of the project team
- How the procurement process will be integrated with other project processes
- Where to find standard procurement documents (provided your organization uses standard documents)
- How many vendors or contractors are involved and how they'll be managed
- How the procurement process will be coordinated with other project processes, such as performance reporting and scheduling
- How the constraints and assumptions might be impacted by purchasing
- How multiple vendors or contractors will be managed
- Coordination of purchasing lead times with the development of the project schedule
- Schedule dates that are determined in each contract
- Identification of prequalified sellers (if known)
- Risk management issues
- Procurement metrics for managing contracts and for evaluating sellers

**Procurement Statements of Work** A procurement statement of work (SOW) contains the details of the procurement item in clear, concise terms. It includes the following elements:

- Project objectives
- Description of the work of the project and any post-project operational support needed
- Concise specifications of the product or services required
- Project schedule, time period of services, and work location

The procurement SOW is prepared by the buyer and is developed from the project scope statement and the WBS and WBS dictionary. The seller uses the SOW to determine whether they are able to produce the goods or services as specified. It will undergo progressive elaboration as the procurement processes occur.

**Make-or-Buy Decisions** The make-or-buy decision is a document that outlines the decisions made during the process regarding which goods and/or services will be produced by the organization and which will be purchased. This can include any number of items:

- Services
- Products
- Insurance policies
- Performance
- Performance bonds

**Procurement Documents** Procurement documents are used to solicit vendors and suppliers to bid on procurement needs. These documents are prepared by the buyer to assure as accurate and complete a response as possible from all potential bidders. Procurement documents have commonly been referred to by the following terms:

- Request for proposal (RFP)
- Request for information (RFI)
- Invitation for bid (IFB)
- Request for quotation (RFQ)

Procurement documents typically include the following information:

- Clear description of the work requested
- Contract SOW
- Explanation of how sellers should format and submit their responses
- Any special provisions or contractual needs

The following words are used during the procurement process:

- When the decision is going to be made primarily on price, the words *bid* and *quotation* are used, as in IFB or RFQ.
- When considerations other than price (such as technology or specific approaches to the project) are the deciding factor, the word *proposal* is used, as in RFP.

Procurement documents are posted or advertised according to the buyer's organizational policies. This might include ads in newspapers and magazines or ads/posts on the Internet.

**Source Selection Criteria** The term *source selection criteria* refers to the method the buyer's organization will use to choose a vendor from among the proposals received. The criteria might be subjective or objective.

Here are some examples of criteria used:

- Purchase price
- Scoring models
- Rating models

The following list includes some of the criteria for evaluating proposals and bids:

- Comprehension and understanding of the needs of the project as documented in the contract SOW
- Cost
- Technical ability of the vendor and their proposed team
- Technical approach
- Risk
- Experience on projects of similar size and scope, including references
- Project management approach
- Management approach
- Financial stability and capacity
- Production capacity
- Reputation, references, and past performance
- Intellectual and proprietary rights

**Change Requests** As a result of going through this process, changes to the project management plan may be required due to vendor capability, availability, cost, quality considerations, and so on. Change requests must be processed through the Perform Integrated Change Control process.

## Exam Essentials

**Be able to define the purpose of the Plan Procurements process.** The purpose of the Plan Procurements process is to identify which project needs should be obtained from outside the organization. Make-or-buy analysis is used as a tool and technique to help determine this.

**Be able to identify the contract types and their usage.** Contract types are a tool and technique of the Plan Procurements process and include fixed-price and cost-reimbursable contracts. Use fixed-price contracts for well-defined projects with a high value to the company, and use cost-reimbursable contracts for projects with uncertainty and large investments early in the project life. The three types of fixed-price contracts are FFP, FPIF, and FP-EPA. The three types of cost-reimbursable contracts are CPFF, CPIF, and CPAF. Time and material contracts are a cross between fixed-price and cost-reimbursable contracts.

**Be able to name the outputs of the Plan Procurements process.** The outputs of Plan Procurements are a procurement management plan, procurement statements of work, make-or-buy decisions, procurement documents, source selection criteria, and change requests.

# Developing a Quality Management Plan

Preventing the occurrence of defects and reducing the cost of quality can be critical to a project's success. To this end, a quality management plan is created out of the Plan Quality process and is responsible for guiding the project management team in carrying out the three quality-related processes. This plan should be based on the project scope and requirements.

In addition to creating the quality management plan, the Plan Quality process is also responsible for creating the process improvement plan. This plan measures the effectiveness of and improves the project management processes. Continuous improvement is a recurring theme throughout the *PMBOK® Guide* and is an important element of modern quality theories. The ideas, concepts, and theories from key quality theorists (such as Deming, Juran, and Crosby) played an important role in shaping the Project Quality Management Knowledge Area processes.

The Plan Quality process is concerned with targeting quality standards that are relevant to the project at hand and devising a plan to meet and satisfy those standards. The quality management plan is the result of the Plan Quality process, which describes how the quality policy will be implemented by the project management team. The result of this process also produces the process improvement plan, which documents the actions for analyzing processes to ultimately increase customer value.

Figure 3.28 shows the inputs, tools and techniques, and outputs of the Plan Quality process.

**FIGURE 3.28** Plan Quality process

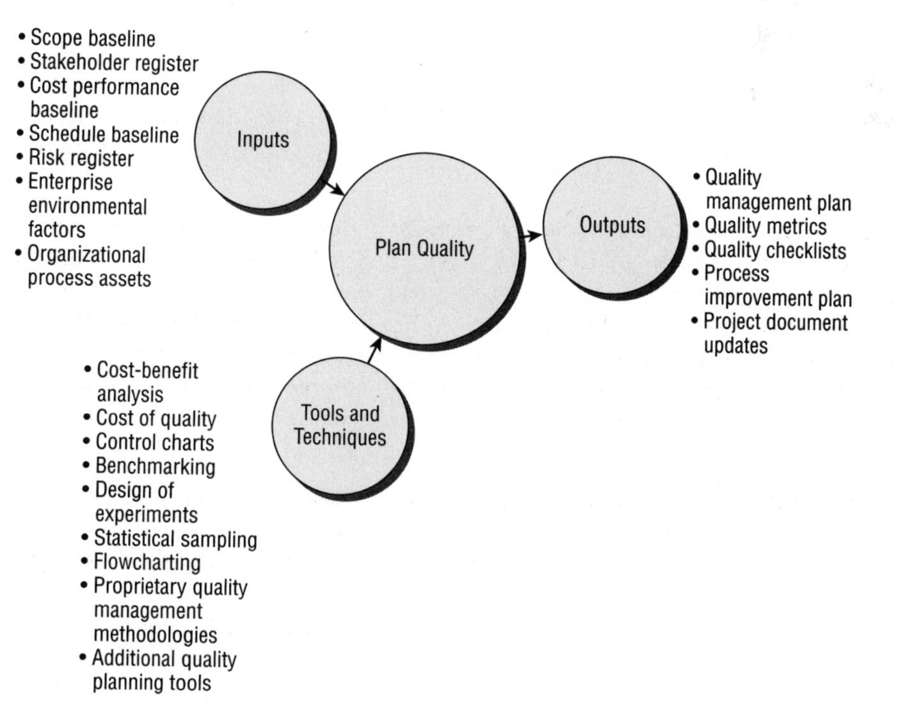

For more detailed information on the Plan Quality process, see Chapter 7 of *PMP: Project Management Professional Exam Study Guide, 6th Edition*.

## Inputs of Plan Quality

The Plan Quality process contains the following inputs:

- Scope baseline
- Stakeholder register
- Cost performance baseline
- Schedule baseline
- Risk register
- Enterprise environmental factors
- Organizational process assets

**Scope Baseline** The following are utilized from within the scope baseline:

- Project scope statement, which defines the project deliverables, objectives, and threshold and acceptance criteria, all of which are used within the quality processes
- WBS
- WBS dictionary

**Stakeholder Register** A list of stakeholders is utilized within this process to understand the expectations of all stakeholders. Quality has been achieved when the expectations of all stakeholders have been met.

**Cost Performance Baseline** Meeting cost performance goals is part of project quality management, and therefore, the cost performance baseline will be utilized within this process.

**Schedule Baseline** Completing activities on schedule is part of the project, as opposed to product, quality management, and therefore, the schedule baseline will be utilized in this process.

**Risk Register** The risk register outlines the documented information pertaining to risk, including the list of risks. Management of risk is tied into quality and should therefore be considered in the Plan Quality process.

**Enterprise Environmental Factors** The project manager should consider any standards, regulations, guidelines, quality policies, or rules that exist concerning the work of the project when writing the quality plan.

**Organizational Process Assets** The quality policy is included from within the organizational process assets and used in this process. The quality policy is a guideline published by executive management that describes what quality policies should be adopted for projects the company undertakes.

**Standards and Regulations**

A standard is something that's approved by a recognized body and that employs rules, guidelines, or characteristics that should be followed. According to the *PMBOK® Guide*, the Project Quality Management Knowledge Area is designed to be in alignment with the ISO standards.

A regulation is mandatory. Governments or institutions almost always impose regulations, although organizations might also have their own, self-imposed regulations.

## Tools and Techniques of Plan Quality

Know the following tools and techniques of the Plan Quality process:

- Cost-benefit analysis
- Cost of quality
- Control charts
- Benchmarking
- Design of experiments
- Statistical sampling
- Flowcharting
- Proprietary quality management methodologies
- Additional quality planning tools

**Cost-Benefit Analysis** In the case of quality management, cost of quality trade-offs should be considered from within cost-benefit analysis. The benefits of meeting quality requirements are as follows:

- Stakeholder satisfaction is increased.
- Costs are lower.
- Productivity is higher.
- There is less rework.

**Cost of Quality** The cost of quality (COQ) is the total cost to produce the product or service of the project according to the quality standards. Three costs are associated with the cost of quality:

**Prevention Costs** Prevention costs are the costs associated with satisfying customer requirements by producing a product without defects.

**Appraisal Costs** Appraisal costs are the costs expended to examine the product or process and make certain the requirements are being met.

**Failure Costs** Failure costs are what it costs when things don't go according to plan. Failure costs are also known as cost of poor quality. Two types of failure costs exist:

- Internal failure costs
- External failure costs

There are two categories of costs within COQ, as listed in Table 3.5.

**TABLE 3.5** Cost of conformance and nonconformance

| Conformance Costs | Nonconformance Costs |
|---|---|
| Prevention costs | Internal failure costs |
| Appraisal costs | External failure costs |

Quality must be planned into the project, not inspected after the fact, to ensure that the product or service meets stakeholders' expectations.

Quality theorists and quality techniques are responsible for the rise of the quality management movement and the theories behind the cost of quality. Figure 3.29 highlights four quality theorists, along with their ideas, that you should be very familiar with for the exam.

**FIGURE 3.29** Quality theorists

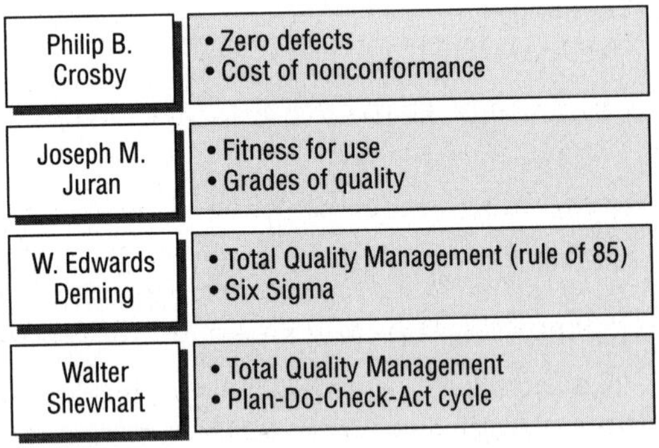

| | |
|---|---|
| Philip B. Crosby | • Zero defects<br>• Cost of nonconformance |
| Joseph M. Juran | • Fitness for use<br>• Grades of quality |
| W. Edwards Deming | • Total Quality Management (rule of 85)<br>• Six Sigma |
| Walter Shewhart | • Total Quality Management<br>• Plan-Do-Check-Act cycle |

**Philip B. Crosby** Philip B. Crosby is known for devising the zero defects practice, which means to do it right the first time. If the defect is prevented from occurring in the first place, costs are lower, conformance to requirements is easily met, and the cost measurement for quality becomes the cost of nonconformance rather than the cost of rework.

**Joseph M. Juran** Joseph M. Juran is known for the fitness for use premise, which means the stakeholders' and customers' expectations are met or exceeded and reflects their views of quality. Juran proposed that there could be grades of quality.

*Grade* is a category for products or services that are of the same type but have differing technical characteristics. *Quality* describes how well the product or service (or characteristics of the product or service) fulfills the requirements. *Low quality* is usually not an acceptable condition, while *low grade* might be.

**W. Edwards Deming** W. Edwards Deming suggested that as much as 85 percent of the cost of quality is management's responsibility, which came to be known as the 85 percent rule or rule of 85. He believed that workers need to be shown what acceptable quality is and that they need to be provided with the right training so that quality and continuous improvement become natural elements of the working environment. Deming was also a major contributor to the modern quality movement and theories, which emphasizes a more proactive approach to quality management (prevention over inspection). For example, he documented 14 points that summarize how management can transform their organization to one that is effective. This and his body of work have been credited for launching the Total Quality Management (TQM) movement. He also popularized the Plan-Do-Check-Act cycle, also referred to as the Deming Cycle, which focuses on continuous improvement.

Six Sigma is a measurement-based strategy that focuses on process improvement and variation reduction by applying Six Sigma methodologies to the project. There are two Six Sigma methodologies:

- DMADV (define, measure, analyze, design, and verify) is used to develop new processes or products at the Six Sigma level.
- DMAIC (define, measure, analyze, improve, and control) is used to improve existing processes or products.

**Walter Shewhart** According to some sources, Walter Shewhart is the grandfather of statistical quality control and the Plan-Do-Check-Act model, which was further popularized by Deming. Shewhart developed statistical tools to examine when a corrective action must be applied to a process. He is also known for the control chart techniques.

**Kaizen Approach** The Kaizen approach—***kaizen*** means improvement in Japanese—is a technique in which all project team members and managers should be constantly watching for quality improvement opportunities. The Kaizen approach states that the quality of the people should be improved first and then the quality of the products or service.

**Control Charts** Control charts help determine if a process is stable and whether process variances are in control or out of control.

**Benchmarking** Benchmarking is a process of comparing previous similar activities to the current project activities to provide a standard against which to measure performance.

**Design of Experiments** Design of experiments (DOE) is a statistical technique attributed to Genichi Taguchi that identifies the elements—or variables—that will have the greatest effect on overall project outcomes. DOE provides a statistical framework that allows the variables that have the greatest effect on overall project outcomes to be changed at once instead of one at a time.

**Statistical Sampling** Statistical sampling involves taking a sample number of parts from the whole population and inspecting them to determine whether they fall within acceptable variances.

**Flowcharting** Flowcharts show the relationships between the process steps. In regard to quality planning, flowcharting helps the project team identify quality issues before they occur.

**Proprietary Quality Management Methodologies** The following are examples of proprietary quality management methodologies:

- Six Sigma
- Lean Six Sigma
- Total Quality Management (TQM)
- Quality Function Deployment

**Additional Quality Planning Tools** The *PMBOK® Guide* lists additional tools, as described in Table 3.6.

**TABLE 3.6** Quality planning tools

| Tool | Description |
| --- | --- |
| Brainstorming | Used to generate ideas within a large group. A brainstorming session can also utilize and review information obtained using the nominal group technique. |
| Affinity diagrams | Used to group and organize thoughts and facts; can be used in conjunction with brainstorming. |
| Force field analysis | A method of examining the drive and resistance of change. |
| Nominal group techniques | Brainstorming sessions consisting of small groups. The ideas of these sessions are later reviewed by a larger group. |
| Matrix diagrams | Used as a decision-making tool, particularly when several options or alternatives are available. |
| Prioritization matrices | Used to prioritize complex issues that have numerous criteria for decision making. |

## Outputs of Plan Quality

There are five outputs of the Plan Quality process:

- Quality management plan
- Quality metrics
- Quality checklists
- Process improvement plan
- Project document updates

**Quality Management Plan** The project manager in cooperation with the project staff writes the quality management plan. The plan should be based on the project scope and requirements to successfully prevent defects from occurring, therefore reducing the cost of quality.

The quality management plan includes the following elements:

- Description of how the project management team will carry out the quality policy
- Resources needed to carry out the quality plan
- Responsibilities of the project team in implementing quality
- All the processes and procedures the project team and organization should use to satisfy quality requirements, including the following items:
  - Quality control
  - Quality assurance techniques
  - Continuous improvement processes

**Quality Metrics** A quality metric, also known as operational definition, describes what is being measured and how it will be measured during the Perform Quality Control process.

**Quality Checklists** Checklists provide a means to determine whether the required steps in a process have been followed. As each step is completed, it's checked off the list.

**Process Improvement Plan** The process improvement plan focuses on finding inefficiencies in a process or activity and eliminating them. The following elements are among those included in the process improvement plan:

- Process boundaries, which describe the purpose for the process and its expected start and end dates
- Process configuration so that you know what processes are performed when and how they interact
- Process metrics
- Any specific elements that should be targeted for improvement

**Project Document Updates** The following project documents may need to be updated:

- Stakeholder register
- RAM

- Quality management plan
- Process improvement plan

**Exam Essentials**

**Be able to identify the benefits of meeting quality requirements.** The benefits of meeting quality requirements include increased stakeholder satisfaction, lower costs, higher productivity, and less rework, and they are discovered during the Plan Quality process.

**Be able to define the cost of quality.** The COQ is the total cost to produce the product or service of the project according to the quality standards. These costs include all the work necessary to meet the product requirements for quality. The three costs associated with cost of quality are prevention, appraisal, and failure costs (also known as cost of poor quality).

**Be able to name four people associated with COQ and some of the techniques they helped establish.** They are Crosby, Juran, Deming, and Shewhart. Some of the techniques they helped to establish are TQM, Six Sigma, cost of quality, and continuous improvement. The Kaizen approach concerns continuous improvement and specifies that people should be improved first.

**Be able to name the tools and techniques of the Plan Quality process.** The Plan Quality process consists of cost-benefit analysis, cost of quality, control charts, benchmarking, design of experiments, statistical sampling, flowcharting, proprietary quality management methodologies, and additional quality planning tools.

# Developing a Change Management Plan

Managing changes is important to successfully managing a project overall. Scope creep is often the result of a poorly documented scope and poorly documented change control procedures. To ensure that changes are managed appropriately, a change management plan should be created. It should be noted that there isn't a formal documented process that creates this plan.

The change management plan should document the following:

- How changes will be monitored and controlled
- Detailed processes for managing change to a project
- How change requests are to be documented and managed
- Process for approving changes
- How to document and manage the final recommendation for the change requests

# Developing a Risk Management Plan

The first official step in performing risk management is creating a risk management plan. This is done by carrying out the Plan Risk Management process, which is the first official process of the Project Risk Management Knowledge Area. Five of the six processes that fall in this knowledge area belong to the Planning process group.

The risk management plan is created to manage uncertainty throughout the project life cycle, which allows the project management team to control the outcome of the project to the extent possible. Like other plans, it plays an important role in guiding the rest of the processes that fall within its respective knowledge area. Without the risk management plan, a project management team cannot effectively identify and manage risks because this plan ensures that everyone is on the same page when it comes to assessing, prioritizing, responding to, and managing risks.

Aside from the Plan Risk Management process, the following risk-related processes fall within the Planning process group: Identify Risks, Perform Qualitative Risk Analysis, Perform Quantitative Risk Analysis, and Plan Risk Responses.

## Plan Risk Management

The Plan Risk Management process determines how the project team will conduct risk management activities. The risk management plan, which will be developed during this process, guides the project team in carrying out the risk management processes and also assures that the appropriate amount of resources and time is dedicated to risk management. By the end of this process, the project team will have developed a common understanding for evaluating risks throughout the remainder of the project.

Figure 3.30 shows the inputs, tools and techniques, and outputs of the Plan Risk Management process.

**FIGURE 3.30** Plan Risk Management process

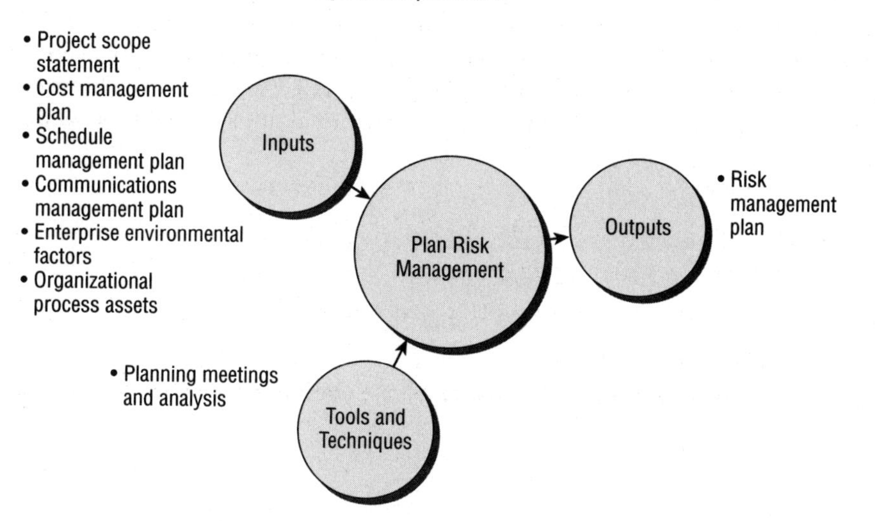

For more detailed information on the Plan Risk Management process, see Chapter 6, "Risk Planning," of *PMP: Project Management Professional Exam Study Guide, 6th Edition*.

## Inputs of Plan Risk Management

You should know the inputs of the Plan Risk Management process:

- Project scope statement
- Cost management plan
- Schedule management plan
- Communications management plan
- Enterprise environmental factors
- Organizational process assets

You should know how each of these inputs relates to the Risk Management process.

**Project Scope Statement** The project scope statement contains the project deliverables, which will be the first stop when identifying risks and determining the process used to evaluate risks.

**Cost Management Plan, Schedule Management Plan, Communications Management Plan** The cost management plan, schedule management plan, and communications management plan all provide important details and information to consider when planning risk management. This includes guidelines for setting aside contingency reserves, information needed to determine risk-reporting formats, and other information important for defining risk activities.

**Enterprise Environmental Factors** One of the key elements of the enterprise environmental factors to consider within this process is the *risk tolerance* levels of the organization and the stakeholders. This is important for evaluating and ranking risk.

**Organizational Process Assets** Organizational process assets include policies and guidelines that might already exist in the organization. The following should be considered when developing the risk management plan:

- Risk categories
- Risk statement formats
- Risk management templates tailored to the needs of the project
- Defined roles and responsibilities
- Authority levels of the stakeholders and project manager

## Tools and Techniques of Plan Risk Management

The Plan Risk Management process has only one tool and technique: planning meetings and analysis.

The purpose of the meetings is to contribute to the risk management plan. During these meetings, the fundamental plans for performing risk management activities will be discussed and determined and then documented in the risk management plan.

## Outputs of Plan Risk Management

The output of the Plan Risk Management process is the risk management plan.

The risk management plan describes how you will define, monitor, and control risks throughout the project. It details how risk management processes will be implemented, managed, monitored, and controlled.

According to the *PMBOK® Guide*, the risk management plan should include the following elements:

**Methodology** Methodology is a description of how you'll perform risk management, including elements such as methods, tools, and where you might find risk data that you can use in the later processes.

**Roles and Responsibilities** The risk management plan should include descriptions of the people who are responsible for managing the identified risks and their responses and the people responsible for each type of activity identified in the plan.

**Budgeting** The budget for risk management activities is included in the plan as well. In this section, you'll assign resources and estimate the costs of risk management activities and its methods. These costs are then included in the cost performance baseline.

**Timing** Timing refers to how often and at what point in the project life cycle risk management processes will occur. You may also include protocols to establish contingency schedule reserves.

**Revised Stakeholder Tolerances** As the project proceeds through the risk management processes, risk tolerances will change. These are documented in the risk management plan. Risk management should be carried out according to stakeholder risk tolerances.

**Reporting formats** In the reporting formats section, you'll describe the content and format of the risk register. You should also detail how risk management information will be maintained, updated, analyzed, and reported to project participants.

**Tracking** This includes a description of how you'll document the history of the risk activities for the current project and how the risk processes will be audited.

**Risk Categories** Risk categories are a way of systematically identifying risks and providing a foundation for understanding. Risk categories should be identified during this process and documented in the risk management plan.

There are multiple ways of displaying risk categories, such as through a simple list or by constructing a risk breakdown structure (RBS), which lists the categories and subcategories. Figure 3.31 shows a sample of an RBS.

**FIGURE 3.31** Risk breakdown structure

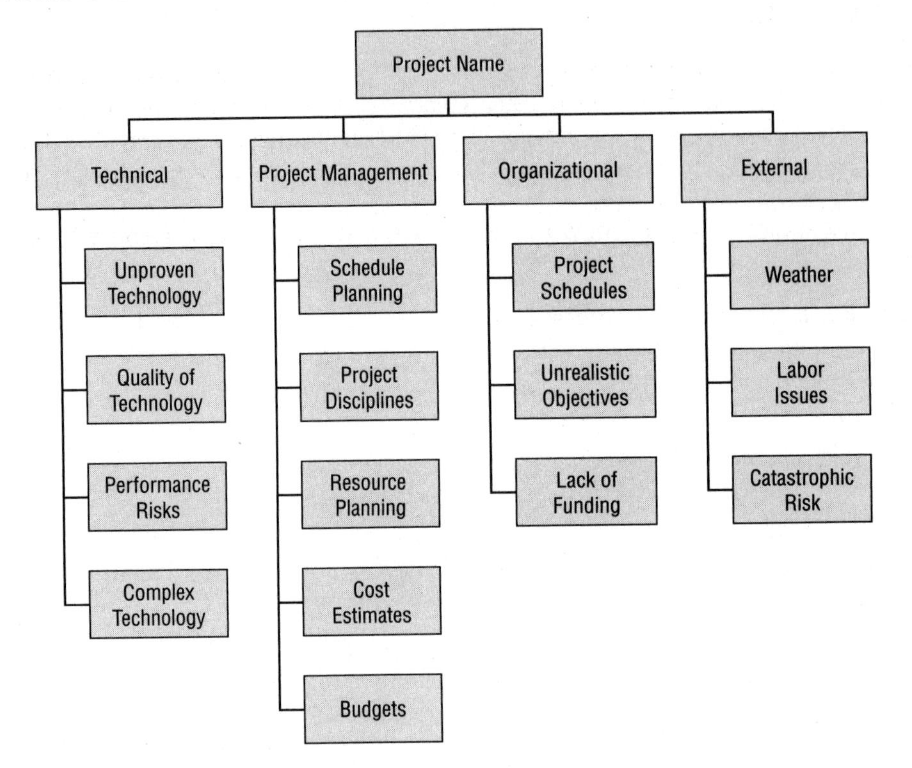


The following list includes some examples of the categories you might consider during this process:

- Technical/quality/performance risks
- Project management risks
- Organizational risks
- External risks

**Defining Probability and Impact** The definitions for probability and impact are documented in the risk management plan as they relate to potential positive or negative risk events and their impacts on project objectives.

- *Probability* describes the potential for the risk event occurring.
- *Impact* describes the effects or consequences the project will experience if the risk event occurs.

**Probability and Impact Matrix** A probability and impact matrix prioritizes the combination of probability and impact scores and helps determine which risks need detailed risk response plans.

## Identify Risks

The Identify Risks process describes all the risks that might impact the project and documents their characteristics in the risk register. Identify Risks is an iterative process that continually builds on itself as additional risks emerge. The risk register is created at the end of this process. The risk register documents all risk information, proposed responses, responses implemented, and status.

Figure 3.32 shows the inputs, tools and techniques, and outputs of the Identify Risks process.

**FIGURE 3.32** Identify Risks process

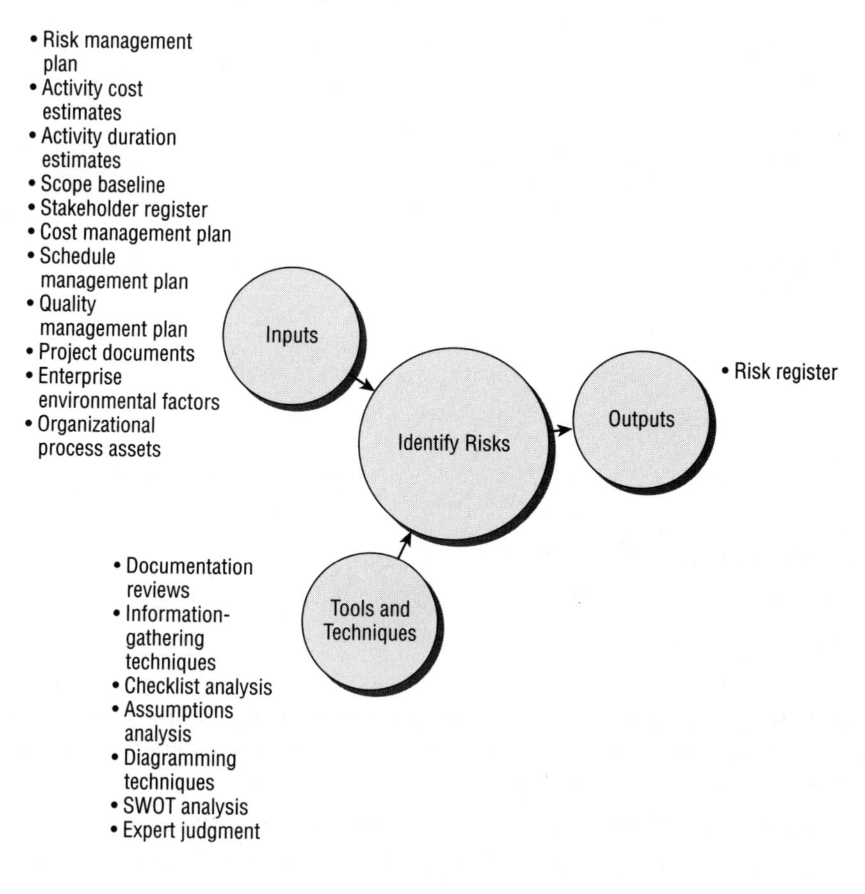

For more detailed information on the Identify Risks process, see Chapter 6 of *PMP: Project Management Professional Exam Study Guide, 6th Edition*.

## Inputs of Identify Risks

You should be familiar with the following inputs of the Identify Risks process:

- Risk management plan
- Activity cost estimates
- Activity duration estimates
- Scope baseline
- Stakeholder register
- Cost management plan
- Schedule management plan
- Quality management plan
- Project documents
- Enterprise environmental factors
- Organizational process assets

**Risk Management Plan** Within this process, the roles and responsibilities section of the risk management plan will be utilized, as well as the budget and schedule set aside for risk activities.

**Activity Cost Estimates** Activity cost estimates may provide insight into the identification of risks if the estimates are found to be insufficient to complete the activity.

**Activity Duration Estimates** Activity duration estimates may reveal existing risks in relation to the time that has been allotted for completing an activity or the project as a whole.

**Scope Baseline** The project scope statement, part of the scope baseline, contains a list of project assumptions, which are issues believed to be true. The assumptions about delivery times are reexamined, and it is determined if they are still valid.

**Stakeholder Register** Understanding stakeholder influence is essential to risk management. The stakeholder register provides a list of stakeholders and may list levels of influence.

**Cost Management Plan, Schedule Management Plan, Quality Management Plan, Project Documents** An understanding of the project management plans is necessary for identifying risks. A thorough review of these documents should be carried out to identify risks associated with them.

**Enterprise Environmental Factors** The enterprise environmental factors utilized include aspects from outside the project that might help determine or influence project outcomes.

Industry information or academic research that might exist for your application areas regarding risk information is especially relevant.

**Organizational Process Assets** Historical information, such as from previous project experiences and project team knowledge, are utilized from within the organizational process assets. Risk register templates can also be used from past similar projects.

## Tools and Techniques of Identify Risks

The Identify Risks process includes the following tools and techniques:

- Documentation reviews
- Information-gathering techniques
- Checklist analysis
- Assumptions analysis
- Diagramming techniques
- SWOT analysis
- Expert judgment

**Documentation Reviews** Documentation reviews involve reviewing project plans, assumptions, and historical information from previous projects from a total project perspective as well as an individual deliverables and activities level. This review helps the project team identify risks associated with the project objectives.

**Information-Gathering Techniques** The following information-gathering techniques are utilized to assist in compiling a comprehensive list of risks:

- Brainstorming
- Delphi technique
- Interviewing
- Root cause analysis

**Checklist Analysis** Checklists used during the Identify Risks process are usually developed based on historical information and previous project team experience. The lowest level of the RBS may be used or a list of risks from previous similar projects. The WBS can also be used as a checklist.

**Assumptions Analysis** Assumptions analysis is a matter of validating the assumptions identified and documented during the course of the project planning processes. Assumptions should be accurate, complete, and consistent. All assumptions are tested against two factors:

- Strength of the assumption or the validity of the assumption
- Consequences that might impact the project if the assumption turns out to be false

**Diagramming Techniques** Three types of diagramming techniques are used in the Identify Risks process:

**Cause-and-Effect** Cause-and-effect diagrams show the relationship between the effects of problems and their causes. This diagram depicts every potential cause and subcause of a problem and the effect that each proposed solution will have on the problem. This diagram is also called a fishbone diagram or Ishikawa diagram after its developer, Kaoru Ishikawa. Figure 3.33 shows a sample cause-and-effect diagram.

**FIGURE 3.33** Cause-and-effect diagram

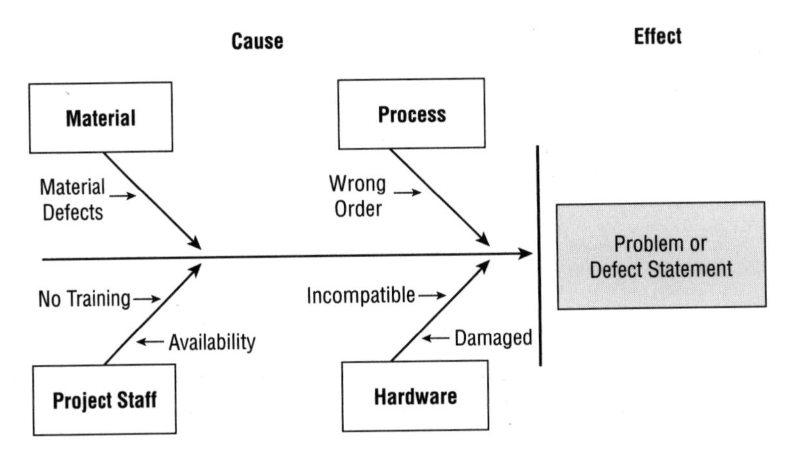


**Process Flowcharts** The system of process flowcharts shows the logical steps needed to accomplish an objective, how the elements of a system relate to each other, and which actions cause which responses. Figure 3.34 shows a flowchart to help determine whether risk response plans should be developed for the risk.

**Influence Diagramming** According to the *PMBOK® Guide*, influence diagramming typically shows the causal influences among project variables, the timing or time ordering of events, and the relationships among other project variables and their outcomes. Figure 3.35 shows an influence diagram for a product introduction decision.

**SWOT Analysis** Strengths, weaknesses, opportunities, and threats (SWOT) analysis is a technique that examines the project from each of these viewpoints and from the viewpoint of the project itself, project management processes, resources, the organization, and so on to identify risks to the project, including risks that are generated internally. SWOT analysis uncovers the following:

- Negative risks, which are typically associated with the organization's weaknesses
- Positive risks, which are typically associated with the organization's strengths

SWOT analysis also determines whether any of the organization's strengths can be used to overcome its weaknesses.

**FIGURE 3.34** Flowchart diagram

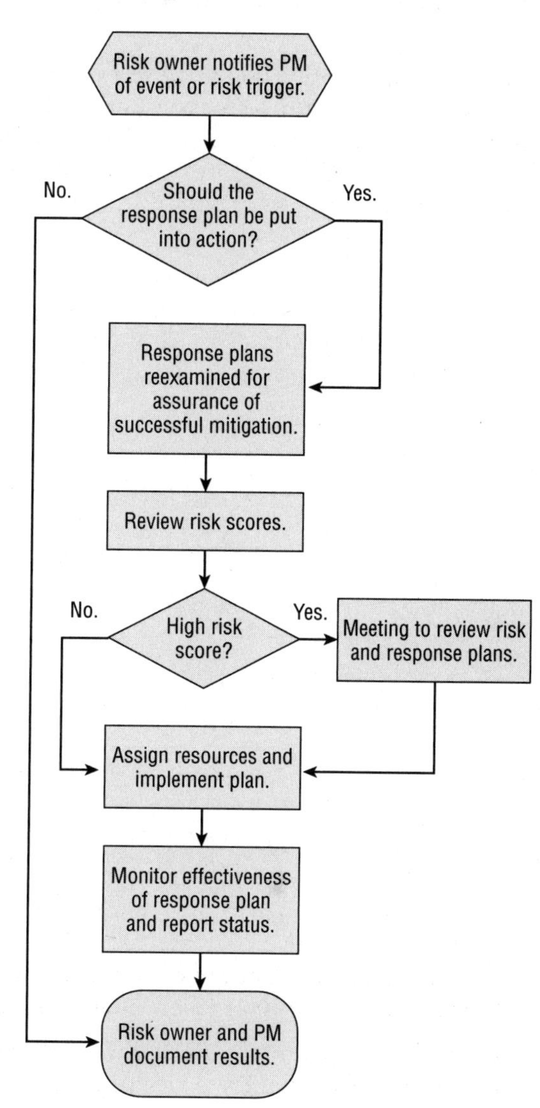


**FIGURE 3.35** Influence diagram

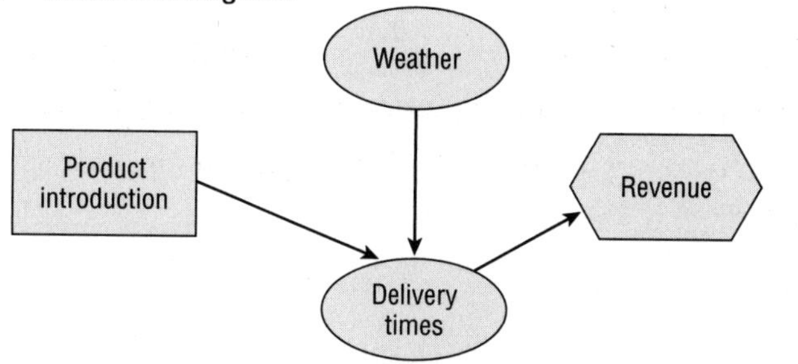

**Expert Judgment** Experts for risk identification purposes can include individuals with the following experience:

- Experience with similar projects
- Experience in the business area for which the project was undertaken
- Industry-specific experience

The bias of experts regarding the project or potential risk events should be considered when using this technique.

## Outputs of Identify Risks

There is only one output of the Identify Risks process: the risk register.

The risk register contains the following elements:

- List of identified risks
- List of potential responses
- Triggers

**List of Identified Risks** Risks are all the potential events and their subsequent consequences that could occur as identified during this process. A list provides a means of tracking risks and their occurrence and responses. The list should contain the following items:

- All potential risks
- Tracking number
- Potential cause or event
- Potential impact
- Responses implemented

**List of Potential Responses** While you're identifying risks, you may identify a potential response at the same time. If this is the case, these potential responses are documented.

**Triggers** Triggers are signals that a risk event is about to occur. Although triggers are not listed as a risk register element, the risk register is an appropriate place to list them in practice.

Table 3.7 shows a sample template for a risk register.

**TABLE 3.7** Risk register template

| ID | Risk | Trigger | Event | Cause | Impact | Owner | Response Plan |
|---|---|---|---|---|---|---|---|
| | | | | | | | |

## Perform Qualitative Risk Analysis

The Perform Qualitative Risk Analysis process involves determining what impact the identified risks will have on the project objectives and the probability that they will occur. It also ranks the risks in priority order according to their effect on the project objectives. The purpose of this process is to determine risk event probability and risk impact. By the end of this process, updates will have been made to the risk register.

Figure 3.36 shows the inputs, tools and techniques, and outputs of the Perform Qualitative Risk Analysis process.

**FIGURE 3.36** Perform Qualitative Risk Analysis process

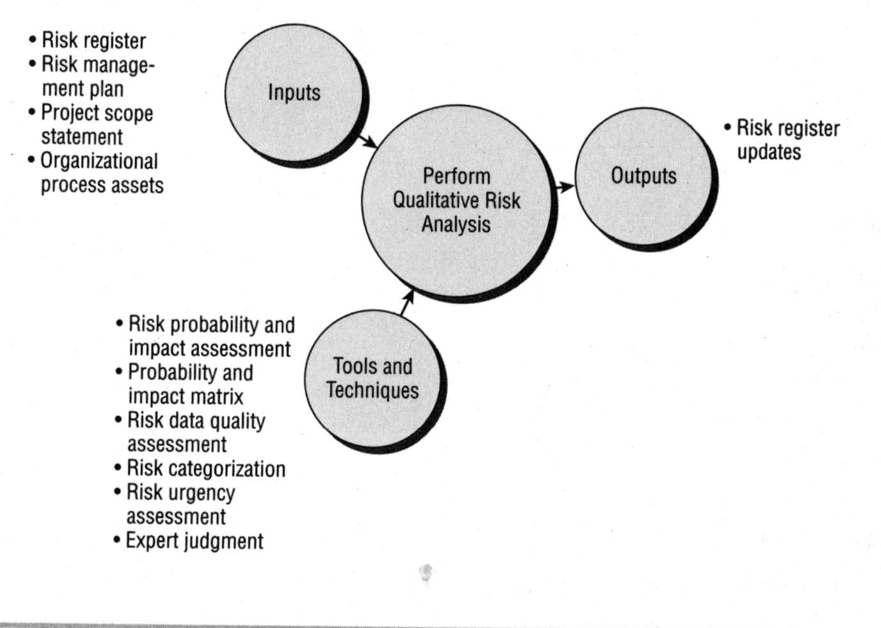

For more detailed information on the Perform Qualitative Risk Analysis process, see Chapter 6 of *PMP: Project Management Professional Exam Study Guide, 6th Edition*.

### Inputs of Perform Qualitative Risk Analysis

Know the following inputs of the Perform Qualitative Risk Analysis process and how they apply:

- Risk register
- Risk management plan
- Project scope statement
- Organizational process assets

**Risk Register** The list of risks from within the risk register is utilized in this process.

**Risk Management Plan** The risk management plan documents the following, which are utilized when prioritizing risks:

- Roles and responsibilities of risk team members
- Budget and schedule factors for risk activities
- Stakeholder risk tolerances
- Definitions for probability and impact
- Probability and impact matrix

**Project Scope Statement** The project scope statement describes the deliverables of the project, which will provide insight into the level of uncertainty that exists within the project.

**Organizational Process Assets** Historical information, lessons learned, and risk databases from past projects are used from within the organizational process assets as a guide for prioritizing the risks for this project.

## Tools and Techniques of Perform Qualitative Risk Analysis

The following tools and techniques are utilized in the Perform Qualitative Risk Analysis process:

- Risk probability and impact assessment
- Probability and impact matrix
- Risk data quality assessment
- Risk categorization
- Risk urgency assessment
- Expert judgment

**Risk Probability and Impact Assessment** This tool and technique assesses the probability that the risk events identified will occur and determines the effect (or impact) they have on the project objectives:

**Probability** Probability is the likelihood that an event will occur.

**Impact** Impact is the amount of pain or gain that the risk event poses to the project. The following are two types of scales used to assign a value to risk:

- The risk impact scale, also known as an ordinal scale, can be a relative scale that assigns values such as high, medium, or low.
- Cardinal scale values are actual numeric values assigned to the risk impact. Cardinal scales are expressed as values from 0.0 to 1.0 and can be stated in equal (linear) or unequal (nonlinear) increments.

Table 3.8 shows a typical risk impact scale for cost, time, and quality objectives based on a high-high to low-low scale.

**TABLE 3.8** Risk impact scale

| Objectives | Low-Low | Low | Medium | High | High-High |
|---|---|---|---|---|---|
| | 0.05 | 0.20 | 0.40 | 0.60 | 0.80 |
| Cost | No significant impact | Less than 6% increase | 7–12% increase | 13–18% increase | More than 18% increase |
| Time | No significant impact | Less than 6% increase | 7–12% increase | 13–18% increase | More than 18% increase |
| Quality | No significant impact | Few components impacted | Significant impact requiring customer approval to proceed | Unacceptable quality | Product not usable |

**Assessing Probability and Impact** The idea behind both probability and impact values is to develop predefined measurements that describe which value to place on a risk event. Assumptions used to arrive at these determinations should also be documented within this process.

**Probability and Impact Matrix** The outcome of a probability and impact matrix is an overall risk rating for each of the project's identified risks. The following are characteristics of the probability and impact matrix:

- The combination of probability and impact results in a classification usually expressed as high, medium, or low.
- Values assigned to the risks determine how the Plan Risk Responses process is carried out for the risks later during the risk-planning processes.
- Values for the probability and impact matrix (and the probability and impact scales) are determined prior to the start of this process.
- The impact and probability matrix is documented in the risk management plan.

In several countries, it is common for high risks to be reflected as a red condition within the probability and impact matrix, medium risks as a yellow condition, and low risks as a green condition. This type of ranking is known as an *ordinal scale* because the values are ordered by rank from high to low.

**Risk Data Quality Assessment** The risk data quality assessment involves determining the usefulness of the data gathered to evaluate risk. The data must be unbiased and accurate. Elements such as the following are used when performing this tool and technique:

- Quality of the data used
- Availability of data regarding the risks
- How well the risk is understood
- Reliability and integrity of the data
- Accuracy of the data

**Risk Categorization** Risk categorization is used to determine the effects risk has on the project.

**Risk Urgency Assessment** Risk urgency assessment determines how soon the potential risks might occur and quickly determines responses for those risks that could occur in the near term. The following play a role in determining how quickly a risk response is needed:

- Risk triggers
- Time to develop and implement a response
- Overall risk rating

**Expert Judgment** Expert judgment is used to determine the probability, impact, and other information derived to date from this process. Interviews and facilitated workshops are two techniques used in conjunction with expert judgment to perform this process.

### Outputs of Perform Qualitative Risk Analysis

The output of the Perform Qualitative Risk Analysis process includes updates to the risk register.

According to the *PMBOK® Guide*, the risk register will receive the following updates, which are added as new entries:

- Risk ranking (or priority) for the identified risks
- Risks grouped by categories
- Causes of risk
- List of risks requiring near-term responses
- List of risks that need additional analysis and response
- Watch list of low-priority risks
- Trends in qualitative risk analysis results

## Perform Quantitative Risk Analysis

The Perform Quantitative Risk Analysis process evaluates the impacts of risk prioritized during the Perform Qualitative Risk Analysis process and quantifies risk exposure for the project by assigning numeric probabilities to each risk and their impacts on project objectives. The purpose of this process is to perform the following:

- Quantify the project's possible outcomes and probabilities.
- Determine the probability of achieving the project objectives.
- Identify risks that need the most attention by quantifying their contribution to overall project risk.
- Identify realistic and achievable schedule, cost, or scope targets.
- Determine the best project management decisions possible when outcomes are uncertain.

Figure 3.37 shows the inputs, tools and techniques, and outputs of the Perform Quantitative Risk Analysis process.

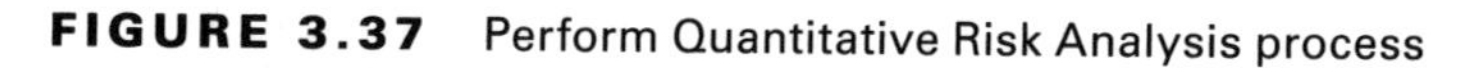
**FIGURE 3.37** Perform Quantitative Risk Analysis process

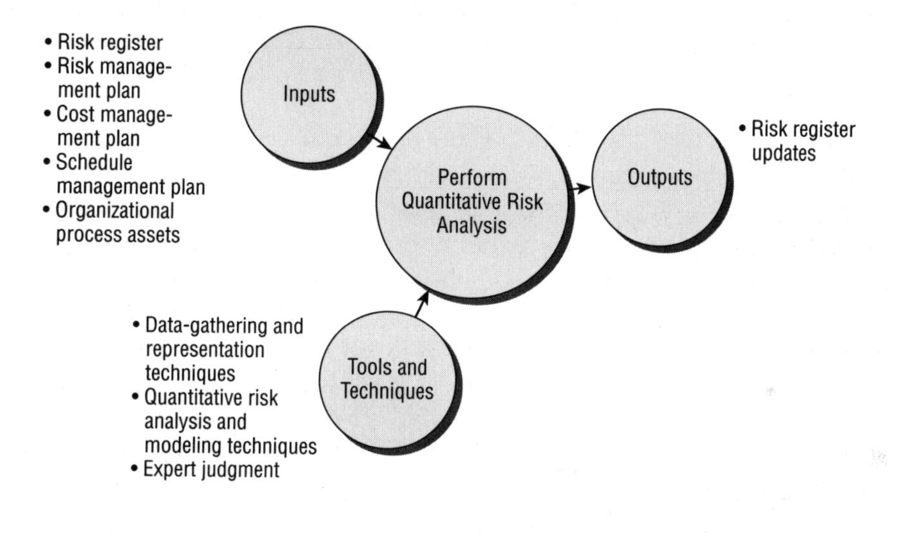

For more detailed information on the Perform Quantitative Risk Analysis process, see Chapter 6 of *PMP: Project Management Professional Exam Study Guide, 6th Edition*.

## Inputs of Perform Quantitative Risk Analysis

Know the following inputs of the Perform Quantitative Risk Analysis process:

- Risk register
- Risk management plan
- Cost management plan

- Schedule management plan
- Organizational process assets

**Risk Register** The risk register contains a list of identified risks as well as any other documented information on risks to date.

**Risk Management Plan** The risk management plan guides the project team in carrying out the risk processes. The following key elements are used from within the risk management plan:

  - Roles and responsibilities
  - Risk management activities
  - Guidelines for setting aside contingency reserves
  - RBS
  - Stakeholder tolerances

**Cost Management Plan** The cost management plan provides the necessary information to address cost-related risks so that the project may be kept within budget.

**Schedule Management Plan** The schedule management plan provides the necessary information to address schedule-related risks so that the project may be kept on schedule.

**Organizational Process Assets** The organizational process assets utilized include the following elements:

  - Historical information from previous projects
  - Risk databases
  - Risk specialists' studies performed on similar projects

## Tools and Techniques of Perform Quantitative Risk Analysis

Three tools and techniques are utilized in the Perform Quantitative Risk Analysis process:

- Data-gathering and representation techniques
- Quantitative risk analysis and modeling techniques
- Expert judgment

**Data-Gathering and Representation Techniques** Data-gathering and representative techniques are as follows:

> **Interviewing** Project team members, stakeholders, and subject matter experts are typical interview subjects. Oftentimes, three-point estimates are gathered from the experts to quantify the probability and impact of risks on project objectives. The following interview topics are common:
>
> - Experiences on past projects
> - Experiences working with the types of technology or processes used during this project

**Probability Distributions** Continuous probability distributions (particularly beta and triangular distributions) are commonly used in Perform Quantitative Risk Analysis. According to the *PMBOK® Guide*, the following probability distributions are often used:

- Normal and lognormal
- Triangular
- Beta
- Uniform distributions
- Discrete distributions

**Distributions are graphically displayed and represent both the probability and time or cost elements.**

**Quantitative Risk Analysis and Modeling Techniques** One or more of the following quantitative risk analysis and modeling techniques are used:

**Sensitivity Analysis** Sensitivity analysis is a quantitative method of analyzing the potential impact of risk events on the project and determining which risk event (or events) has the greatest potential for impact by examining all the uncertain elements at their baseline values. Figure 3.38 shows a sample tornado diagram, which is one of the ways that sensitivity analysis data can be displayed.

**FIGURE 3.38** Tornado diagram

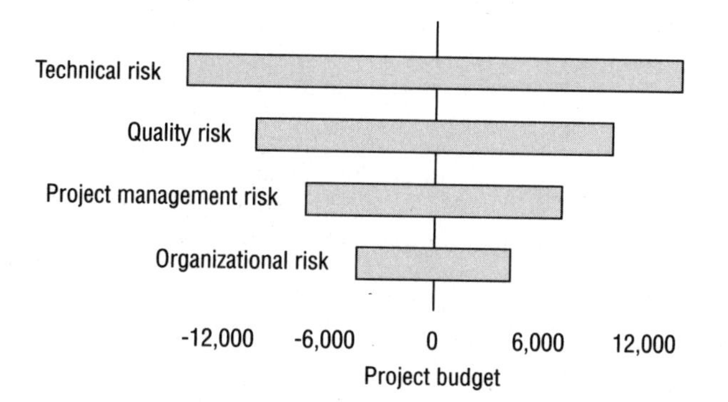


**Expected Monetary Value (EMV) Analysis** Expected monetary value (EMV) analysis is a statistical technique that calculates the average, anticipated future impact of the decision. EMV is calculated by multiplying the probability of the risk by its impact and then adding them together:

- Positive results generally indicate an opportunity within the project.
- Negative results generally indicate a threat to the project.

EMV is used in conjunction with the decision tree analysis technique. Decision trees are diagrams that show the sequence of interrelated decisions and the expected results of choosing one alternative over the other. Typically, more than one choice or option is available in response to a decision. The available choices are depicted in tree form starting at the left with the risk decision branching out to the right with possible outcomes. Decision trees are usually used for risk events associated with time or cost. Figure 3.39 shows a sample decision tree using expected monetary value (EMV) as one of its inputs.

**FIGURE 3.39** Decision tree

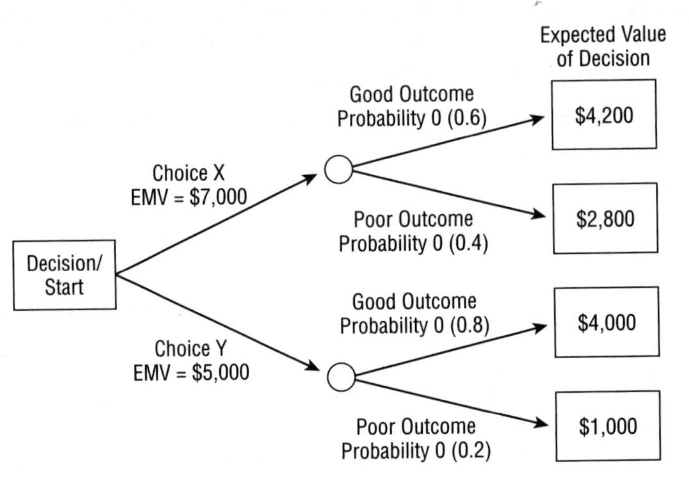


**Modeling and Simulation** Modeling and simulation techniques are often used for schedule risk analysis and cost analysis. Simulation techniques compute the project model using various inputs, such as cost or schedule duration, to determine a probability distribution for the variable chosen. Modeling and simulation techniques examine the identified risks and their potential impacts to the project objectives from the perspective of the whole project.

Monte Carlo analysis is an example of a simulation technique. It is replicated many times, typically using cost or schedule variables. Every time the analysis is performed, the values for the variable are changed with the output plotted as a probability distribution. This type of information helps the risk management team determine the probability of completing the project on time and/or within budget.

**Expert Judgment** Experts can come from inside or outside the organization and should have experience that's applicable to the project. When performing quantitative risk analysis, experts can also assist in validating the data and tools used.

## Outputs of Perform Quantitative Risk Analysis

Like the process before it, the Perform Quantitative Risk Analysis process contains one output, updates to the risk register.

The following risk register updates occur as a result of this process:

- Probabilistic analysis of the project
- Probability of achieving the cost and time objectives
- Prioritized list of quantified risks
- Trends in Perform Quantitative Risk Analysis results

**Probabilistic Analysis of the Project** Probabilistic analysis of the project is the forecasted results of the project schedule and costs as determined by the outcomes of risk analysis. The following are characteristics of probabilistic analysis:

- Results include projected completion dates and costs along with a confidence level associated with each.
- The output is often expressed as a cumulative distribution.
- Results are used along with stakeholder risk tolerances to quantify the time and cost contingency reserves.

**Probability of Achieving the Cost and Time Objectives** The probability of achieving the cost and time objectives of the project are documented based on the results of performing the quantitative risk analysis tools and techniques.

**Prioritized List of Quantified Risks** The prioritized list of risks includes the following items:

- Risks with the greatest risk or threat to the project
- Risk impacts
- Risks that present the greatest opportunities to the project
- Risks that are most likely to impact the critical path
- Risks with the largest cost contingency

**Trends in Perform Quantitative Risk Analysis Results** Trends in Perform Quantitative Risk Analysis appear as the process is repeated. Trends are documented for further analysis and for use in developing risk response plans.

## Plan Risk Responses

Plan Risk Responses is a process of deciding what actions to take to reduce threats and take advantage of the opportunities discovered during the risk analysis processes. This includes assigning resources the responsibility of carrying out risk response plans outlined in this process. By the end of this process, the risk register will be updated to include a risk response plan for all risks that require some form of action.

Figure 3.40 shows the inputs, tools and techniques, and outputs of the Plan Risk Responses process.

**FIGURE 3.40** Plan Risk Responses process

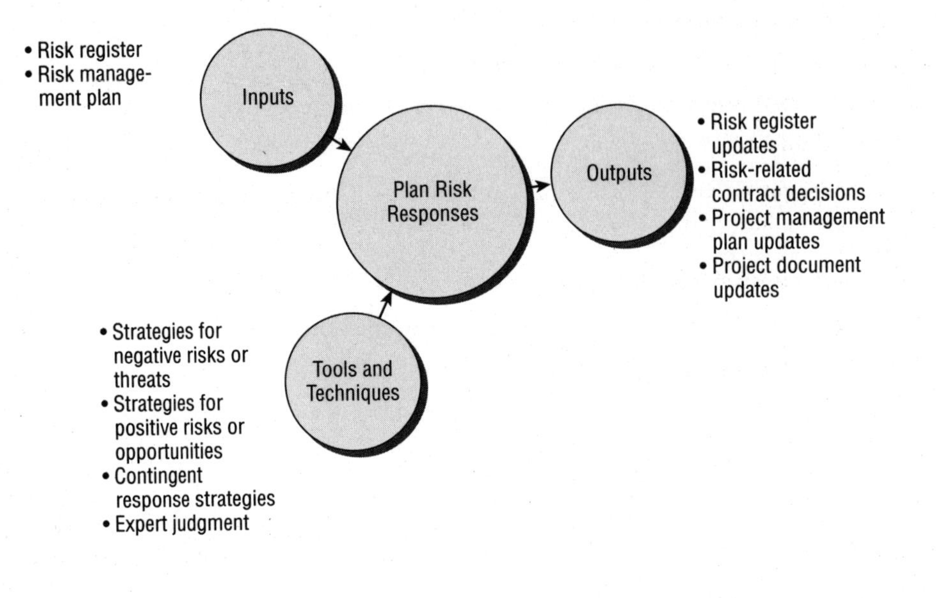

For more detailed information on the Plan Risk Responses process, see Chapter 6 of *PMP: Project Management Professional Exam Study Guide, 6th Edition.*

## Inputs of Plan Risk Responses

The Plan Risk Responses process contains two inputs that you should be familiar with. Be sure you know how they are used during the process.

**Risk Register** All of the information included within the risk register to date will be utilized in developing risk responses.

**Risk Management Plan** The following are utilized from within the risk management plan in carrying out this process:

- Roles and responsibilities that pertain to risk
- Definitions of risk analysis
- Risk thresholds
- Time and budget requirements allotted for risk activities

## Tools and Techniques of Plan Risk Responses

Tools and techniques of the Plan Risk Responses process that you should know are as follows:

- Strategies for negative risks or threats
- Strategies for positive risks or opportunities
- Contingent response strategies
- Expert judgment

**Strategies for Negative Risks or Threats** As Figure 3.41 shows, strategies for negative risks or threats include avoid, transfer, mitigate, and accept.

**FIGURE 3.41** Strategies for negative risks

| Strategies for Negative Risks | • Avoid<br>• Transfer<br>• Mitigate<br>• Accept |
|---|---|

**Avoid** To avoid a risk means to evade it altogether, eliminate the cause of the risk event, or change the project plan to protect the project objectives from the risk event.

**Here are some examples of avoiding risk:**

- Improving communications
- Refining requirements
- Assigning additional resources to project activities
- Refining the project scope to avoid risk events

**Transfer** The purpose of risk transfer is to transfer a risk and its consequences to a third party. This strategy will impact the project budget.

**Here are some examples of transferring risk:**

- Insurance
- Contracting
- Warranties
- Guarantees
- Performance bonds

**Mitigate** To mitigate a risk means to reduce the probability of a risk event and its impacts to an acceptable level. According to the *PMBOK® Guide*, the purpose of mitigation is to reduce the probability that a risk will occur and reduce the impact of the risk to an acceptable level.

**Accept** The acceptance strategy is used when the project team isn't able to eliminate all the threats to the project. Acceptance of a risk event is a strategy that can be used for risks that pose either threats or opportunities to the project. There are two forms of acceptance:

- Passive acceptance occurs when the project team has assessed the risk as low enough in probability and/or impact that they choose to do no additional planning for that potential event at this time.
- Active acceptance includes developing contingency reserves to deal with risks should they occur.

**Strategies for Positive Risks or Opportunities** Strategies for positive risks or opportunities, as shown in Figure 3.42, include exploit, share, enhance, and accept.

**FIGURE 3.42** Strategies for positive risks

| Strategies for Positive Risks | • Exploit<br>• Share<br>• Enhance<br>• Accept |
|---|---|

**Exploit** Exploiting a risk event is to look for opportunities for positive impacts. This is the strategy of choice when positive risks have been identified and the project team wants to make certain the risk will occur within the project. An example of exploiting a risk would be reducing the amount of time to complete the project by bringing on more qualified resources.

**Share** The share strategy is similar to transferring because risk is assigned to a third-party owner who is best able to bring about the opportunity the risk event presents. An example of sharing a risk would be a joint venture.

**Enhance** The enhance strategy involves closely watching the probability or impact of the risk event to assure that the organization realizes the benefits. This entails watching for and emphasizing risk triggers and identifying the root causes of the risk to help enhance impacts or probability.

**Accept** This is similar to the accept strategies used for negative risks or threats.

**Contingent Response Strategies** Contingent response strategies, also known as contingency planning, involves planning alternatives to deal with the risks should they occur. Contingency planning recognizes that risks may occur and that plans should be in place to deal with them.

The following contingency responses are common:

- Contingency reserves, which include project funds that are held in reserve to offset any unavoidable threats that might occur to project scope, schedule, cost, or quality. It also includes reserving time and resources to account for risks.
- Fallback plans, which are developed for risks with high impact or for risks with identified strategies that might not be the most effective at dealing with the risk.

**Expert Judgment** Expert judgment is utilized in developing risk responses, including feedback and guidance from risk management experts and those internal to the project qualified to provide assistance in this process.

## Outputs of Plan Risk Responses

The following are the four outputs of the Plan Risk Responses process:

- Risk register updates
- Risk-related contract decisions
- Project management plan updates
- Project document updates

**Risk Register Updates** According to the *PMBOK® Guide*, after Identify Risks, Perform Qualitative Risk Analysis, and Perform Quantitative Risk Analysis are preformed, the following elements should appear in the risk register:

- List of identified risks, including their descriptions, what WBS element they impact (or area of the project), categories (RBS), root causes, and how the risks impact the project objectives
- Risk owners and their responsibility
- Outputs from the Perform Qualitative Analysis process
- Agreed-upon response strategies
- Actions needed to implement response plans
- Risk triggers
- Cost and schedule activities needed to implement risk responses
- Contingency plans
- Fallback plans, which are risk response plans that are executed when the initial risk response plan proves to be ineffective
- Contingency reserves
- Residual risk, which is a leftover risk that remains after the risk response strategy has been implemented
- Secondary risks, which are risks that come about as a result of implementing a risk response

**Risk-Related Contract Decisions** If risk response strategies include responses such as transference or sharing, it may be necessary to purchase services or items from third parties. Contracts for those services can be prepared and discussed at this time.

**Project Management Plan Updates** The following documents within the project management plan may require updates:

- Any and all of the management plans
- WBS

- Schedule baseline
- Cost performance baseline

**Project Document Updates** Other project documents, such as technical documentation and the assumptions documented in the project scope statement, may require an update after performing this process.

## Exam Essentials

**Be able to define the purpose of the risk management plan.** The risk management plan describes how you will define, monitor, and control risks throughout the project. It details how risk management processes (including Identify Risks, Perform Qualitative Risk Analysis, Perform Quantitative Risk Analysis, Plan Risk Responses, and Monitor and Control Risks) will be implemented, monitored, and controlled throughout the life of the project. It describes how you will manage risks but does not attempt to define responses to individual risks. The risk management plan is a subsidiary of the project management plan, and it's the only output of the Plan Risk Management process.

**Be able to define the purpose of the Identify Risks process.** The purpose of the Identify Risks process is to identify all risks that might impact the project, document them, and identify their characteristics.

**Be able to define the risk register and some of its primary elements.** The risk register is an output of the Identify Risks process, and updates to the risk register occur as an output of every risk process that follows this one. By the end of the Plan Risk Responses process, the risk register contains these primary elements: identified list of risks, risk owners, risk triggers, risk strategies, contingency plans, and contingency reserves.

**Be able to define the purpose of the Perform Qualitative Risk Analysis process.** Perform Qualitative Risk Analysis determines the impact the identified risks will have on the project and the probability they'll occur, and it puts the risks in priority order according to their effects on the project objectives.

**Be able to define the purpose of the Perform Quantitative Risk Analysis process.** Perform Quantitative Risk Analysis evaluates the impacts of risks prioritized during the Perform Qualitative Risk Analysis process and quantifies risk exposure for the project by assigning numeric probabilities to each risk and their impacts on project objectives.

**Be able to define the purpose of the Plan Risk Responses process.** Plan Risk Responses is the process where risk response plans are developed using strategies such as avoid, transfer, mitigate, accept, exploit, share, enhance, develop contingent response strategies, and apply expert judgment. The risk response plan describes the actions to take should the identified risks occur. It should include all the identified risks, a description of the risks, how they'll impact the project objectives, and the people assigned to manage the risk responses.

# Obtaining Project Management Plan Approval

With the subsidiary plans and baselines created, it's time to bring them all together into a project management plan. The project management plan consists of a compilation of these plans and baselines and is an output of the Develop Project Management Plan process. This primary plan is presented to key stakeholders for sign-off (if required). Once it's approved, the team can move forward to execute the plan.

Develop Project Management Plan is the first process within the Planning process group, and it's part of the Project Integration Management Knowledge Area. This process defines, coordinates, and integrates all the various subsidiary project plans and baselines. The result is a project management plan document that describes how the project outcomes will be executed, monitored, and controlled as the project progresses and how the project will be closed out once it concludes.

Throughout the project, the project management plan will be updated to reflect approved project changes. It will be used as an input to many of the project management processes, and as a result of approved changes, it will also become an output of several processes.

Figure 3.43 shows the inputs, tools and techniques, and outputs of the Develop Project Management Plan process.

**FIGURE 3.43** Develop Project Management Plan process

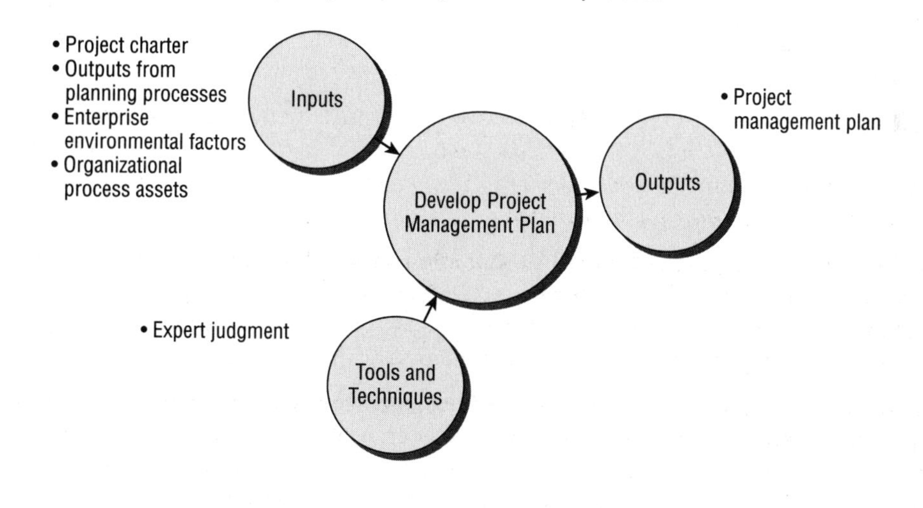

For more detailed information on the Develop Project Management Plan process, see Chapter 3 of *PMP: Project Management Professional Exam Study Guide, 6th Edition.*

## Inputs of Develop Project Management Plan

Know the following inputs of the Develop Project Management Plan process:

- Project charter
- Outputs from the planning processes
- Enterprise environmental factors
- Organizational process assets

**Project Charter** Based on the high-level scope outlined in the project charter, the project team can start determining which project management processes will be most beneficial for this particular project.

**Outputs from Planning Processes** The following outputs from the planning processes may be helpful inputs to develop the project management plan:

- Any processes used that produce a baseline
- Any processes that produce a subsidiary management plan

**Enterprise Environmental Factors** The following key elements of the environmental factors should be considered when choosing the processes to perform for this project:

- Standards and regulations (both industry and governmental)
- Company culture and organizational structure
- Personnel administration
- Project management information system (PMIS)

The PMIS is an automated or manual system used to document the subsidiary plans and the project management plan, facilitate the feedback process, and revise the documents.

**Organizational Process Assets** The following key elements of the organizational process assets should be considered within this process:

- Project management plan template
- Change control procedures
- Historical information
- Configuration management knowledge database that contains the official company policies, standards, procedures, and other project documents

## Tools and Techniques of Develop Project Management Plan

Expert judgment is the only tool and technique of the Develop Project Management Plan process.

The following types of expert judgment are needed to complete this process:

- Tailoring techniques
- Understanding technical and management details that need to be included in the project management plan

- Determining resources and assessing skill levels needed for project work
- Determining and defining the amount of configuration management to apply on the project

## Output of Develop Project Management Plan

The output of the Develop Project Management Plan process is the project management plan.

The purpose of most processes is to produce an output. Outputs are usually a report or document of some type or a deliverable.

The project management plan includes the following elements:

- Processes used to perform each phase of the project
- Life cycle used for the project and for each phase of the project when applicable
- Tailoring of results the project team defines
- Methods for executing the work of the project to fulfill the objectives
- Change management plan describing methods for monitoring and controlling change
- Configuration management
- Methods for determining and maintaining the validity of performance baselines
- Communication needs of the stakeholders and techniques to fulfill those needs
- Management reviews of content, issues, and pending decisions

The project management plan also contains multiple subsidiary plans and baselines, as described in Table 3.9. Figure 3.44 depicts a high-level view of the contents that make up the plan.

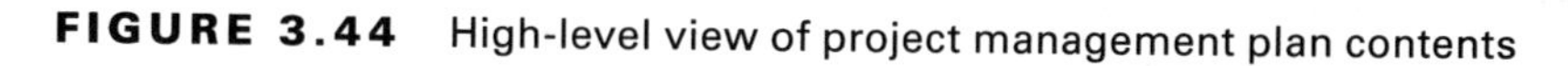

**FIGURE 3.44** High-level view of project management plan contents

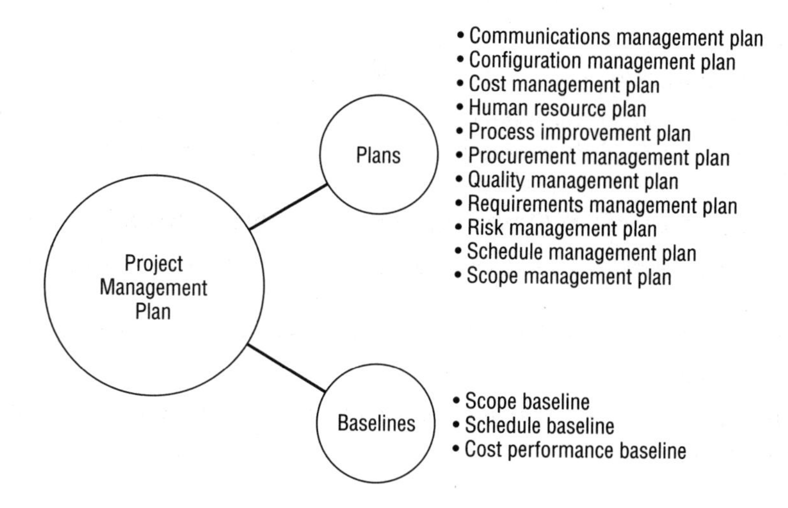

**TABLE 3.9** Project management plan summary

| Subsidiary Plan or Document | Description |
|---|---|
| Requirements management plan | This plan documents how the project requirements will be managed throughout the life of the project. It also addresses how the requirements will be analyzed and documented. |
| Schedule management plan | This plan identifies the scheduling format, tool, and methodology that will be used for creating and managing the project schedule. |
| Cost management plan | This plan defines the criteria for planning, budgeting, estimating, and managing the costs of the project. It also identifies the cost-related formats and tools that will be used within the project. |
| Quality management plan | This plan describes how the organization's quality policies will be implemented and addresses the following items: quality assurance, quality control, and process improvement. |
| Communications management plan | This plan guides the communication for the project. It defines the communication requirements of the stakeholders as well as the format, type, frequency, and methods for project reports. It also lists report recipients and anything else pertinent to project communication. |
| Risk management plan | This plan defines how risk management activities will be carried out. It also outlines the budget and time allotted for risk activities, risk roles and responsibilities, risk categories, and how risks will be defined and assessed. |
| Procurement management plan | This plan describes how procurements will be managed, from make-or-buy decisions all the way through contract closure. |
| Schedule baseline | This baseline is an approved version of the project schedule that will be used to compare and measure the planned schedule against the actual schedule. |
| Cost performance baseline | This baseline contains an approved budget that is used to measure, monitor, and control the planned budget against the actual amount of funds used. |
| Scope baseline | This baseline provides scope boundaries for the project and consists of the project scope statement, the work breakdown structure (WBS), and the WBS dictionary. |

Subsidiary and component plans may be detailed or simply a synopsis, depending on the project needs. As the project progresses and more and more processes are performed, the subsidiary plans and the project management plan itself may change.

**Exam Essentials**

**Be able to state the purpose of the Develop Project Management Plan process.** It defines how the project is executed, monitored and controlled, and closed, and it documents the processes used during the project.

# Conducting a Kickoff Meeting

A kickoff meeting typically occurs at the end of the planning process, prior to beginning the project work. The purpose of the kickoff meeting is to ensure that everyone is aware of the project details and their role within the project and to announce that the project work is ready to begin. This is a great opportunity to review project milestones and other important information with key stakeholders.

## Meeting Attendees

Those who attend a kickoff meeting include, but are not limited to, the following individuals:

- Project manager
- Project team members
- Customer
- Project sponsor
- Department managers
- Other key stakeholders

## Meeting Topics

The topics discussed within a kickoff meeting include, but are not limited to, the following:

- Introductions of attendees
- Meeting agenda
- Review of the project management plan

# Bringing the Processes Together

This chapter covered a tremendous amount of information; the Planning process group is the largest project management process group. In addition to understanding each process individually, you'll need to understand how the processes work together. Understanding that bigger picture will help you with situational exam questions as well as utilizing the project management processes in a real-world setting.

Now that you have a strong grasp of how the outputs from various processes become the inputs for other processes, you understand why some processes must occur before others can move forward. For example, to create the project management plan, the results of the initiating process group are required. Figure 3.45 shows this interaction between the Initiating process group and the Planning process group, the first two stages of the project's life cycle. The relationship between the project management plan and the rest of the planning processes is just as interactive, if not more dynamic.

**FIGURE 3.45** Planning Process Group interaction

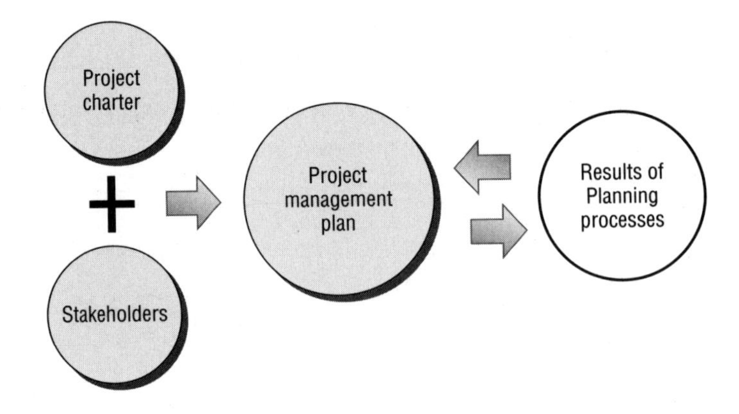


The project management plan is used in planning processes across eight of the nine Knowledge Areas. (The remaining Knowledge Area generates the project management plan.) The results of these planning processes are fed back into the project management plan as updates. The updates might, in turn, impact several of the other planning processes. For example, the project management plan contains the cost management plan. Any changes that are approved within the cost management plan become updates to the project management plan. As a result, these approved and updated changes might impact other planning processes that might *also* require that updates and changes be made.

By now, it should be clear that the project management plan is the heart of the project and that much of the project's success can be based on the processes of the Planning process group. A common motto within the project management industry is, "A well-planned project makes for a successful project." Figure 3.46 illustrates how the project management plan remains at the center of the planning process group.

**FIGURE 3.46** Planning process group triangle

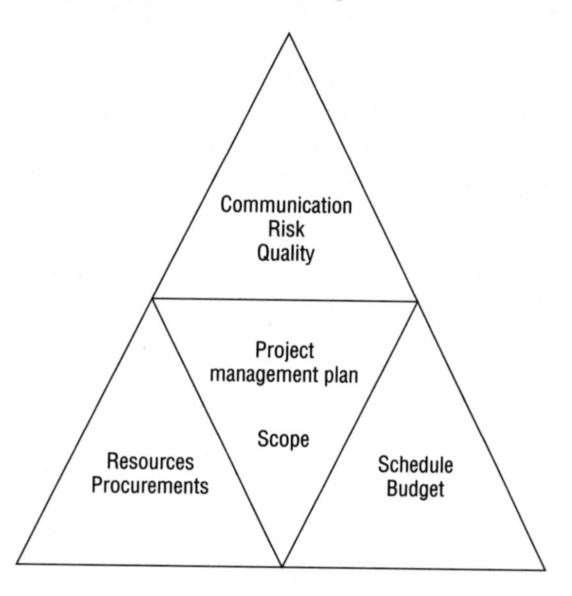


In summary, the project management plan is a compilation of several subsidiary plans and baselines:

- Change management plan
- Communications management plan
- Configuration management plan
- Cost management plan
- Cost performance baseline
- Human resource plan
- Process improvement plan
- Procurement management plan
- Quality management plan
- Requirements management plan
- Risk management plan
- Schedule baseline
- Schedule management plan
- Scope baseline
- Scope management plan

The change management plan, which was not previously addressed, describes how you will document and manage change requests, the process for approving changes, and how to document and manage the final recommendation for the change requests. Configuration management changes deal with the components of the product of the project, such as functional ability or physical attributes, rather than the project process itself.

Scope is also a critical element within planning because it defines the boundaries of the project and what it is that the project has set out to achieve. As shown in Figure 3.46, the project schedule and the project budget are part of the project's foundation. These project elements play a big role in measuring and monitoring the progress of the project. The resources and purchased services and/or products complete the project's foundation. Without the human resources, as well as the external project teams, the project could not move forward.

Guiding the project from above are communications management, quality management, and risk management—all critical elements of the project that influence and guide the project toward success. As you can see, all of these project elements are intertwined and build one on another.

With this big picture in mind, let's go back and review what you've learned thus far about each of the project management Knowledge Areas. You'll find that what goes on within each Knowledge Area is connected on a high level and is structured in a logical format.

## Project Scope Management Knowledge Area Review

You may recall that the project scope defines the work to be completed during the project, providing the project with boundaries. Figure 3.47 illustrates the process and inputs used to create the work breakdown structure. (Remember that the WBS is a decomposition of the project deliverables, down to the work package level.) The WBS becomes part of the scope baseline, as is the project scope statement and the WBS dictionary. The scope baseline is necessary input for activities in the next Knowledge Area, Project Time Management.

**FIGURE 3.47** Project Scope Management Knowledge Area process interaction

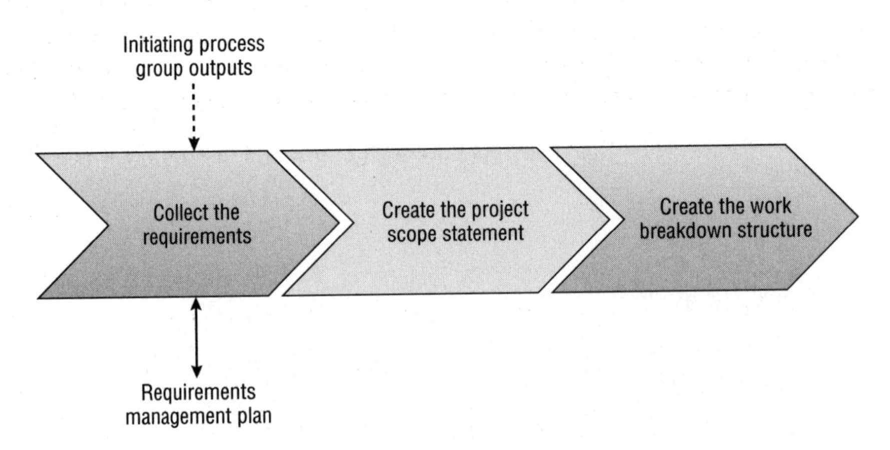

## Project Time Management Knowledge Area Review

Figure 3.48 reflects, step by step, how the project schedule is developed. The planning process flow within the Project Time Management Knowledge Area is extremely clear. It's important to understand the following points:

- The scope baseline is central to creating the initial list of project activities; creating the list is the first step toward developing the project schedule.
- The project budget can influence the number and type of resources the project can use.
- Clearly defining the resources needed to carry out the project activities is key to planning human resources and procurements.

**FIGURE 3.48** Project Time Management Knowledge Area process interaction

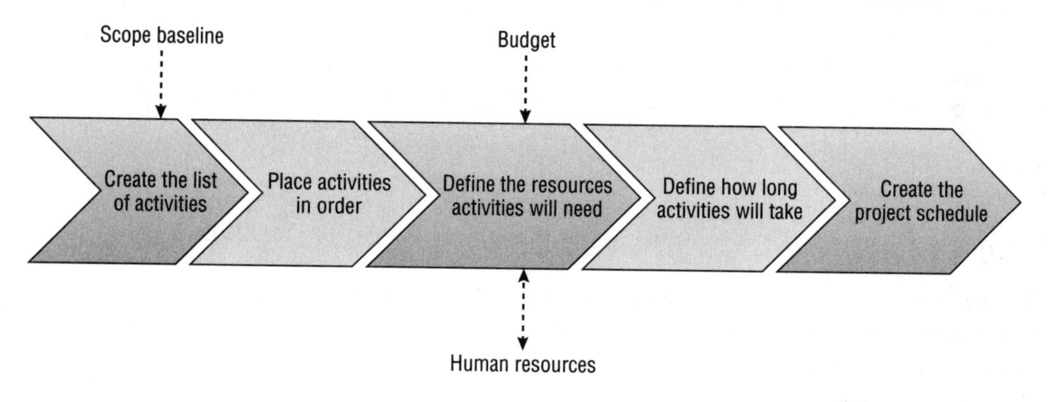


## Project Cost Management Knowledge Area Review

Figure 3.49 shows the key process steps for creating the project budget. The scope baseline is necessary for determining estimates for the cost of each activity. Although not shown as an output of the first cost-related process, the cost management plan comes into play and governs how the activity costs are estimated and how the project budget is created.

**FIGURE 3.49** Project Cost Management Knowledge Area process interaction

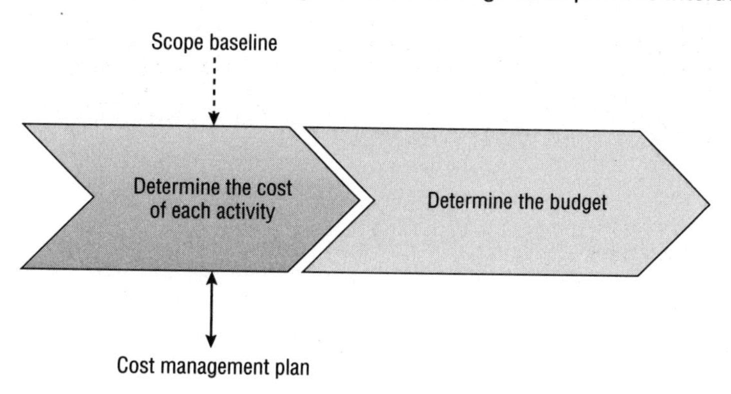

## Project Quality Management Knowledge Area Review

The single planning step within the Project Quality Management Knowledge Area results in several key items within the quality management processes. Figure 3.50 shows how the information resulting from scope, cost, time, and risk planning processes is necessary for carefully planning project quality. This first step in defining project quality produces the following:

- Method for planning, carrying out, and controlling quality activities
- Metrics used to measure project quality
- Method the project management team can use to approach and then carry out a process improvement strategy

**FIGURE 3.50** Project Quality Management Knowledge Area process interaction

## Project Human Resource Management Knowledge Area Review

Figure 3.51 depicts the only planning process within the Project Human Resource Management Knowledge Area. Key to this first process of the Knowledge Area is developing the human resource management plan, which will guide the rest of the processes relating to human resources, including hiring, developing, and managing the project team.

## Project Communications Management Knowledge Area Review

As you learned in this chapter, meeting stakeholder needs, requirements, and expectations is essential to a successful project. Project communications play a big role in meeting these objectives. Figure 3.52 depicts the only planning-related step within the Project Communications Management Knowledge Area. This step determines how project communications will be managed and defines stakeholder communication requirements. The result is the communications management plan. To define the communication needs and requirements of stakeholders, you need the following results of the Initiating process group:

- Stakeholder register
- Stakeholder management strategy

**FIGURE 3.51** Project Human Resource Management Knowledge Area process interaction

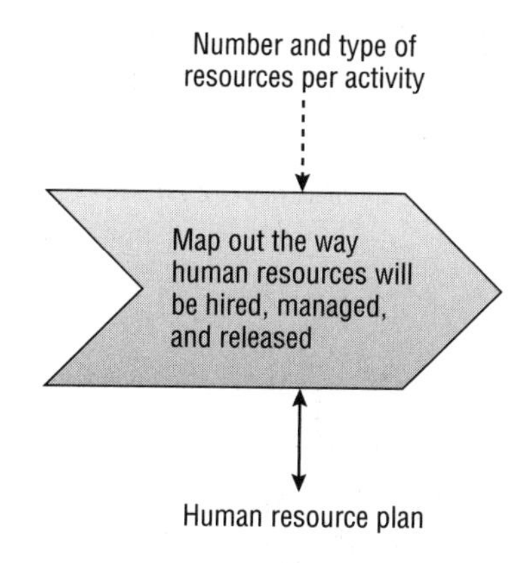


**FIGURE 3.52** Project Communications Knowledge Area process interaction

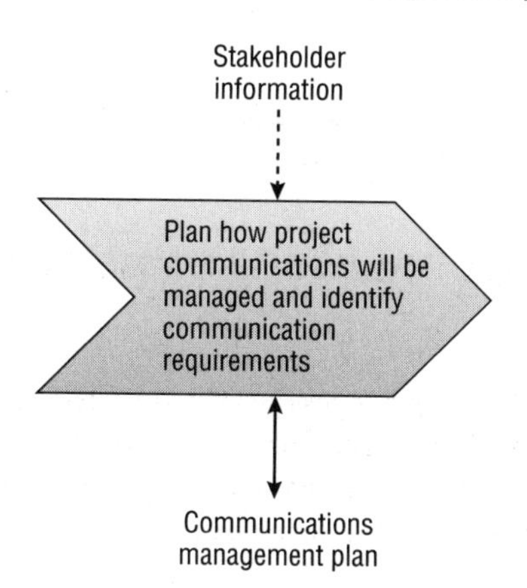


## Project Risk Management Knowledge Area Review

Risk management planning begins as early as the project begins. The first step, which creates the risk management plan, is carried through early on during the planning phase of the project. Project plans, such as the cost management plan, the schedule management plan, and the communications management plan, are used to develop the risk management plan. The project scope statement is also utilized.

As depicted in Figure 3.53, the steps within risk management planning are very intuitive. First, you plan how to manage, assess, and deal with risks. Then, you identify and prioritize the identified risks. Analyzing the risks comes next. As you may recall, this can include qualitative risk analysis and quantitative risk analysis.

**FIGURE 3.53** Project Risk Management Knowledge Area process interaction

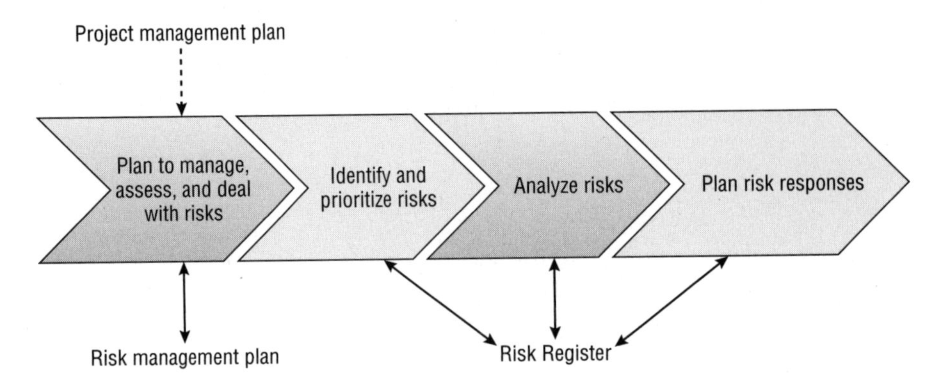


With the existing information at hand, you can now plan how you will respond to risks that call for action. As you can see in Figure 3.53, the risk register is a living document, updated throughout the project as risks are identified. It documents all of the information about these risks and is therefore continuously updated as the processes are carried out.

## Project Procurement Management Knowledge Area Review

Figure 3.54 shows the single procurement-planning step. Here you plan how procurements will be carried out and managed within the project. This activity also determines which products or services will be procured. A total of 11 inputs feed the process.

The results include the following items:

- A plan that lays out the guidelines for carrying out procurement activities, the types of contracts that will be used, and how vendor progress and performance will be measured
- Make-or-buy decisions
- Several key procurement-related documents, including procurement terms, the procurement statement of work, and source selection criteria

As you can see, the planning processes within each Knowledge Area are highly interactive. The processes interact not just within their own respective Knowledge Areas, but among the other planning processes.

Each project carries different needs, and not all of the processes will be necessary for every project. We previously touched on the fact that some processes can be combined to tailor management for a particular project to the needs of that individual project. But, as you've also seen, some process outputs are critical as inputs to other processes. The project management plan, which lies at the core of the Planning process group, is one of those critical outputs.

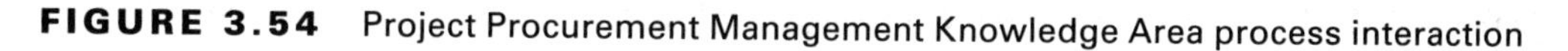

**FIGURE 3.54** Project Procurement Management Knowledge Area process interaction

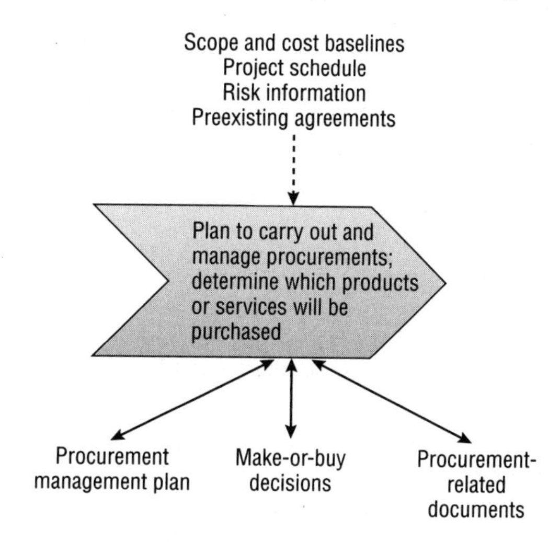

# Review Questions

1. All of the following are subsidiary elements of the project management plan EXCEPT:
   - **A.** Quality management plan
   - **B.** Risk management plan
   - **C.** Project scope statement
   - **D.** Scope baseline

2. The following BEST describes a focus group:
   - **A.** Gathering of prequalified subject matter experts and stakeholders
   - **B.** One-on-one conversations with stakeholders
   - **C.** Group of subject matter experts who participate anonymously
   - **D.** Group of cross-functional stakeholders who can define cross-functional requirements

3. Sam is currently in the planning stages of a project and is working on developing the project scope statement. As part of converting the product description into deliverables, he has utilized the product breakdown and systems analysis techniques. Which of the following tools and techniques of the Define Scope process is Sam currently utilizing?
   - **A.** Expert judgment
   - **B.** Product analysis
   - **C.** Alternatives identification
   - **D.** Facilitated workshops

4. The following are included within the WBS dictionary EXCEPT:
   - **A.** Code of accounts identifiers
   - **B.** Cost estimates
   - **C.** Description of the work of the component
   - **D.** Activity list

5. While sequencing activities, a project manager decides to utilize the precedence diagramming method to show existing dependencies between activities. Which of the logical relationships is the project manager most likely to use?
   - **A.** Finish-to-start
   - **B.** Start-to-finish
   - **C.** Finish-to-finish
   - **D.** Start-to-start

6. Rodrigo is in the Plan Communications process of a new systems component project. He has determined that there are 32 stakeholders within the project. How many communication channels exist?
   A. 496
   B. 512
   C. 32
   D. 31
7. What is the purpose of the Perform Qualitative Risk Analysis process?
   A. To determine how the project team will plan for risks
   B. To identify and gather all potential risks within the project
   C. To determine the impact and probability of identified risks and prioritize them
   D. To evaluate the impact of risks prioritized and determine the risk exposure
8. Which type of contract carries the highest risk for the buyer?
   A. Fixed-price contracts
   B. Lump-sum contracts
   C. Cost-reimbursable contracts
   D. Time and material contracts
9. Which of the following quality theorists devised the zero defects practice?
   A. Walter Shewhart
   B. Philip B. Crosby
   C. W. Edwards Deming
   D. Joseph Juran
10. All of the following are tools and techniques of the Plan Quality process EXCEPT:
    A. Statistical sampling
    B. Flowcharts
    C. Force field analysis
    D. Operational definition

# Answers to Review Questions

1. C. The project scope statement is an output of the Define Scope process and is not included as part of the project management plan. The following plans and components are included: scope management plan, requirements management plan, schedule management plan, cost management plan, quality management plan, communications management plan, risk management plan, procurement management plan, schedule baseline, cost performance baseline, and the scope baseline.

2. A. Option A refers to those participating in a focus group, while B refers to interviews, C refers to the Delphi technique, and D refers to facilitated workshops.

3. B. Sam is currently utilizing product analysis, which is a method for converting the product description and project objectives into deliverables and requirements. To do this, Sam may have utilized any of the following: value analysis, functional analysis, requirements analysis, systems-engineering techniques, systems analysis, product breakdown, and value-engineering techniques.

4. D. The WBS contains deliverables that are broken down to the work package level. The work packages are not broken down into the activity level until the Define Activities process, which cannot occur until after the WBS has been created.

5. A. Finish-to-start is the type of logical relationship most frequently used within precedence diagramming methods and therefore the most likely choice. Option B, start-to-finish, is used the least.

6. A. The formula for calculating communications channels is *n multiplied by n* – 1, all divided by 2, where *n* represents the total number of stakeholders within the project. When you plug in the number of stakeholders in Rodrigo's project, the following formula results:

   $32(32 - 1) \div 2$

7. C. Option A refers to Plan Risk Management, B refers to Identify Risks, and D refers to Perform Quantitative Risk Analysis.

8. C. Cost-reimbursable contracts carry the highest risk for the buyer because the total cost of the goods or services purchased is uncertain and the seller charges all costs to the buyer with no preset cap. As a side note, options A and B refer to the same type of contract.

9. B. Philip B. Crosby believed that preventing defects from occurring resulted in lower costs and conformance to requirements and that cost of quality transformed to cost of conformance as opposed to rework.

10. D. Option D is another name for quality metrics, which is an output of the Plan Quality process. Although C is not directly listed as a tool and technique of this process, it is listed by the *PMBOK® Guide* as an additional quality planning tool.

Chapter

4

# Executing the Project

## THE PMP EXAM CONTENT FROM THE EXECUTING THE PROJECT PERFORMANCE DOMAIN COVERED IN THIS CHAPTER INCLUDES THE FOLLOWING:

- ✓ Obtain and manage project resources, including outsourced deliverables, by following the procurement plan, in order to ensure successful project execution.
- ✓ Maximize team performance through leading, mentoring, training, and motivating team members.
- ✓ Execute the tasks as defined in the project plan, in order to achieve the project deliverables within budget and schedule.
- ✓ Implement approved changes according to the change management plan, in order to meet project requirements.
- ✓ Implement the quality management plan using the appropriate tools and techniques, in order to ensure that work is being performed according to required quality standards.
- ✓ Implement approved actions (e.g., workarounds) by following the risk management plan, in order to minimize the impact of the risks on the project.

Executing is the third process group of the five project management process groups. The processes in this group are responsible for executing the work outlined in the project management plan, managing the project resources, and performing the work of the project. Within this process group, you will also see approved changes and actions implemented as well as re-planning and re-baselining as a result of implemented changes.

Executing also involves keeping the project in line with the original project plan. The majority of the project budget and time will be spent in this process group, and the majority of conflicts will relate to the schedule. In addition, the product description will be finalized and contain more detail than it did in the planning processes.

The Executing process group accounts for 30 percent of the questions on the Project Management Professional (PMP®) exam.

The process names, inputs, tools and techniques, outputs, and descriptions of the project management process groups and related materials and figures in this chapter are based on content from *A Guide to the Project Management Body of Knowledge, 4th Edition (PMBOK® Guide)*.

# Obtaining and Managing Resources

As mentioned previously, the Executing process group involves executing the project work. Naturally, this cannot be accomplished without first acquiring the resources needed to perform the project work. Obtaining resources occurs through the Acquire Project Team process and the Conduct Procurements process. The human resource and procurement management plans play an important role in carrying out these processes because they outline how resources will be acquired, managed, and released throughout the project life cycle.

## Acquire Project Team

The Acquire Project Team process involves acquiring and assigning human resources, both internal and external, to the project. It is the project manager's responsibility to ensure that resources are available and skilled in the project activities to which they're assigned. This process considers the following factors:

- The process of negotiation with individuals who can provide the needed human resources

- The consequences of not obtaining the human resources
- Alternative resources should the planned resources not be available as a result of circumstances out of the project manager's control

Figure 4.1 shows the inputs, tools and techniques, and outputs of the Acquire Project Team process.

**FIGURE 4.1** Acquire Project Team process

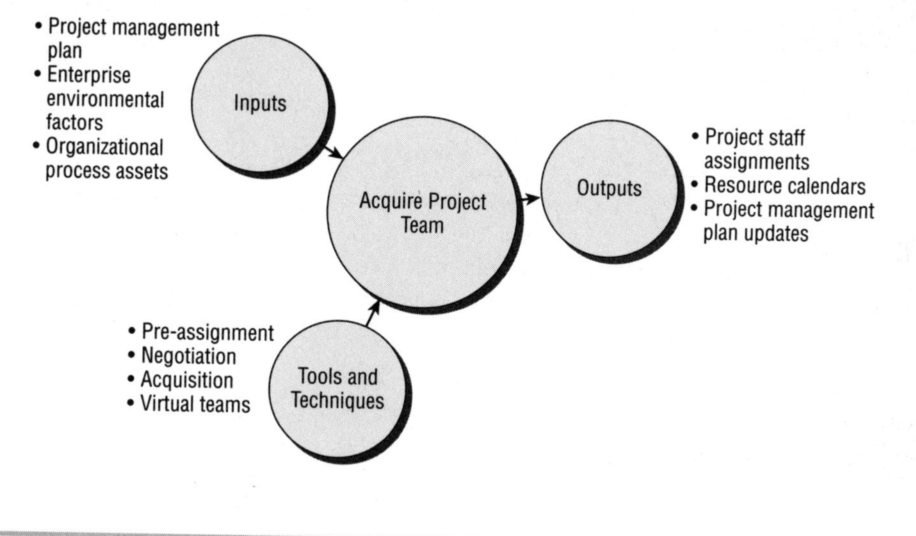

For more detailed information on the Acquire Project Team process, see Chapter 8, "Developing the Project Team," of *PMP: Project Management Professional Exam Study Guide, 6th Edition* (Sybex, 2011).

## Inputs of Acquire Project Team

Know the following inputs of the Acquire Project Team process:

**Project Management Plan** The human resource plan, a subsidiary of the project management plan, is essential to this process. The following are utilized from within this plan:

- Roles and responsibilities
- Project organization chart
- Staffing management plan

The staffing management plan should detail how the team will be acquired.

**Enterprise Environmental Factors** The enterprise environmental factors used in this process involve taking the following information into account before making assignments:

- Personal interests
- Cost rates

- Prior experience
- Availability of potential team members
- Personnel administration policies
- Organizational structure
- Locations

**Organizational Process Assets** The organizational process assets input refers to the standard processes, policies, procedures, and guidelines that the organization has in place. This includes policies around acquiring external resources. In particular, recruitment practices should be taken into account.

## Tools and Techniques of Acquire Project Team

Be familiar with the following tools and techniques of the Acquire Project Team process:

- Pre-assignment
- Negotiation
- Acquisition
- Virtual teams

**Pre-assignment** Pre-assignment may take place when the project is put out for bid and specific team members are promised as part of the proposal or when internal project team members are promised and assigned as a condition of the project. Pre-assignments are documented within the project charter.

**Negotiation** Project managers may use negotiation techniques when dealing with functional managers and other organizational department managers—and sometimes with the vendor to get some of their best people—for project resources and for the timing of those resources.

The following items are typically negotiated:

- Availability
- Competency level of the staff member assigned

**Acquisition** Acquisition involves hiring individuals or teams of people for certain project activities, as either employees or contract help during the course of the project or project phase or for specific project activities.

**Virtual Teams** According to the *PMBOK® Guide*, virtual teams are defined as groups of people with a shared goal who fulfill their roles with little or no time spent meeting face to face. The use of virtual teams makes it possible to draw in resources that wouldn't otherwise be available. It also reduces travel expenses by allowing teams to work from home. Virtual teams typically connect using technology tools, such as the Internet, email, and videoconferencing.

Communication becomes essential when functioning in a virtual structure. All team members should be made aware of the protocols for communicating in a virtual team environment, understand the expectations, and be clear on the decision-making processes.

## Outputs of Acquire Project Team

Know the following three outputs of the Acquire Project Team process:

**Project Staff Assignments** Project staff assignments are based on the results of negotiating and determining elements such as the roles and responsibilities and reviewing recruitment practices. This output also results in a published project team directory, which lists the names of all project team members and stakeholders.

**Resource Calendars** Resource calendars show the team members' availability and the times they are scheduled to work on the project. A composite resource calendar includes availability information for potential resources as well as their capabilities and skills.

**Project Management Plan Updates** As a result of performing the Acquire Project Team process, updates will be made to the human resource plan and staffing management plan.

# Conduct Procurements

In some cases, resources will need to be procured externally. The Conduct Procurements process is concerned with obtaining responses to bids and proposals from potential vendors, selecting a vendor, and awarding the contract. This process is used only when goods or services are obtained from outside of the project's organization. After this process is conducted, sellers will have been selected, contracts awarded, and project documents updated to reflect the selected vendors.

Figure 4.2 shows the inputs, tools and techniques, and outputs of the Conduct Procurements process.

**FIGURE 4.2** Conduct Procurements process

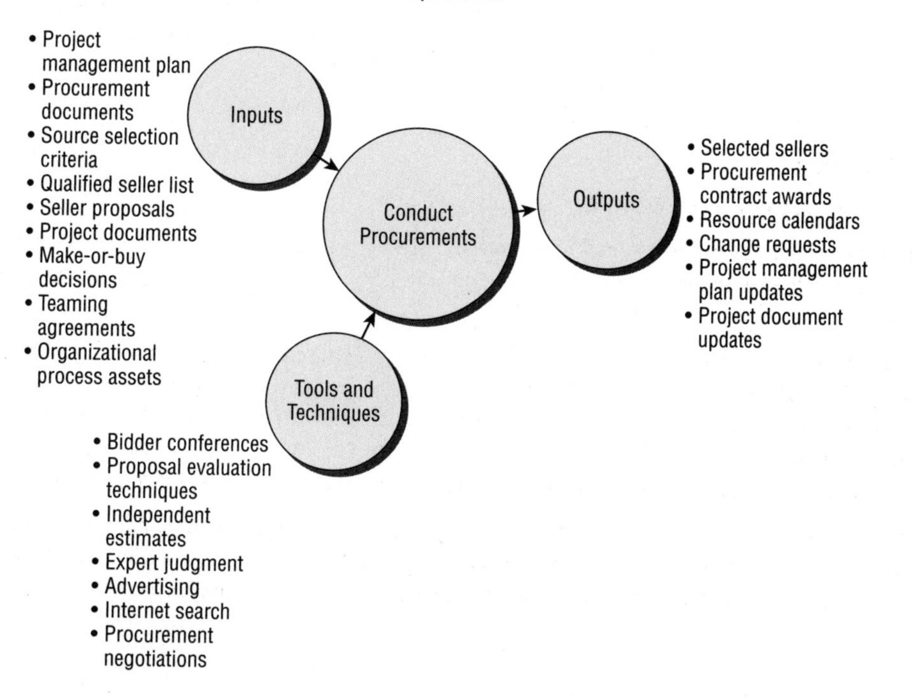

For more detailed information on the Conduct Procurements process, see Chapter 9, "Conducting Procurements and Sharing Information," in *PMP: Project Manager Professional Exam Study Guide, 6th Edition.*

## Inputs of Conduct Procurements

There are several inputs of the Conduct Procurements process that you should know:

- Project management plan
- Procurement documents
- Source selection criteria
- Qualified seller list
- Seller proposals
- Project documents
- Make-or-buy decisions
- Teaming agreements
- Organizational process assets

**Project Management Plan** The procurement management plan is used within the project management plan to guide how the process will be managed.

**Procurement Documents** The following procurement documents are typically utilized:

- Requests for proposals (RFPs)
- Requests for information (RFIs)
- Requests for quotations (RFQs)

**Source Selection Criteria** Source selection criteria includes information on the vendor requirements that the project organization will use to make a decision on which vendor to select.

**Qualified Seller List** Qualified seller lists are lists of prospective sellers who have been preapproved or prequalified to provide contract services (or provide supplies and materials) for the organization.

**Seller Proposals** Seller proposals are literally the proposals submitted by sellers in response to a procurement document package.

**Project Documents** Project documents typically used as an input to this process include the risk register and risk-related contract decisions.

**Make-or-Buy Decisions** Make-or-buy decisions reflect which project deliverables will be purchased and which will be handled internally within the organization. Buy decisions will go through this process to select the vendor.

**Teaming Agreements** Existing teaming agreements are predetermined by executive management. In this process, the buyer will prepare the procurement statement of work, and a contract will be negotiated with the seller.

**Organizational Process Assets** Organizational process assets may contain information on sellers used in past projects. This information is utilized within this process.

## Tools and Techniques of Conduct Procurements

Know the following tools and techniques of the Conduct Procurements process:

- Bidder conferences
- Proposal evaluation techniques
- Independent estimates
- Expert judgment
- Advertising
- Internet search
- Procurement negotiations

**Bidder Conferences** Bidder conferences (also known as vendor conferences), prebid conferences, and contractor conferences are meetings with prospective vendors or sellers that occur prior to the completion of their response proposal. Bidder conferences have the following characteristics:

- They allow all prospective vendors to meet with the buyers to ask questions and clarify issues they have regarding the project and the RFP.
- They are held once.
- They are attended by all vendors at the same time.
- They are held before vendors prepare their responses.

**Proposal Evaluation Techniques** There are several techniques that can be used to evaluate proposals. The types of goods and services you're trying to procure will dictate how detailed your evaluation criteria are. Depending on the complexity of the procurements, you may use one or more of the following techniques to evaluate and rate sellers:

- Use the source selection criteria process to rate and score proposals.
- When purchasing goods, request a sample from each vendor to compare quality (or some other criteria) against your need.
- Ask vendors for references and/or financial records.
- Evaluate the response itself to determine whether the vendor has a clear understanding of what you're asking them to do or provide.

After utilizing the necessary techniques, compare each proposal against the criteria, and rate or score each proposal for its ability to meet or fulfill these criteria. This can serve as your first step in eliminating vendors that don't match your criteria. The next step is to apply the tools and techniques of this process to further evaluate the remaining potential vendors.

Using a weighting system, you can assign numerical weights to evaluation criteria and then multiply them by the weights of each criteria factor to come up with total scores for each vendor. Screening systems and seller rating systems are also sometimes used.

**Independent Estimates** Independent estimates, also known as cost estimates, may be compiled by the buyer's organization to compare to the vendor prices and serve as a benchmark.

**Expert Judgment** Expert judgment may include experts from all areas of the organization when evaluating proposals and selecting vendors.

**Advertising** Advertising lets potential vendors know that an RFP is available. It can be used as a way of expanding the pool of potential vendors, or it may be a requirement, such as in the case of government projects. Here are some examples of where advertising may appear:

- The company's Internet site
- Professional journals
- Newspapers

**Internet Search** Internet searches can be used for multiple actions by the buyer:

- Find vendors
- Perform research on their past performance
- Compare prices
- Purchase items that are readily available and are generally offered for a fixed price

**Procurement Negotiations** In procurement negotiations, both parties come to an agreement regarding the contract terms. At a minimum, a contract should include the following items:

- Price
- Responsibilities
- Applicable regulations or laws that apply
- Overall approach to the project

Once agreement is reached and the negotiations are finished, the contract is signed by both buyer and seller and is executed.

## Outputs of Conduct Procurements

The Conduct Procurements process contains the following six outputs:

- Selected sellers
- Procurement contract awards

- Resource calendars
- Change requests
- Project management plan updates
- Project document updates

**Selected Sellers** Selected sellers are vendors that have been chosen to provide the goods or services requested by the buyer.

**Procurement Contract Awards** A contract is a legally binding agreement between two or more parties, typically used to acquire goods or services. Contracts have several names:

- Agreements
- Memorandums of understanding (MOUs)
- Subcontracts
- Purchase orders

The type of contract awarded depends on the product or services being procured and on the organizational policies.

A contract's life cycle consists of four stages: requirement, requisition, solicitation, and award. In the requirement stage, the project and contract needs are established and requirements are defined within the statement of work (SOW). The requisition stage focuses on refining and confirming the project objectives and generating solicitation materials, such as the request for proposals (RFP), request for information (RFI), and request for quotations (RFQ). The solicitation stage is where vendors are asked to compete and respond to the solicitation materials. And last, the award stage is where the vendors are chosen and contracts awarded.

The contract should clearly address the following:

- Elements of the SOW
- Time period of performance
- Pricing and payment plan
- Acceptance criteria
- Warranty periods
- Dispute resolution procedures and penalties
- Roles and responsibilities

According to the *PMBOK® Guide*, a negotiated draft contract is one of the requirements of the selected sellers output. Also note that senior management signatures may be required on complex, high-risk, or high-dollar contracts. Be certain to check your organization's procurement policies regarding the authority level and amounts you are authorized to sign for.

**Resource Calendars** Resource calendars will contain the following information on procurement resources:

- Quantity
- Availability
- Dates resources are active or idle

**Change Requests** Change requests may include changes to the project management plan and its subsidiary plans and components. These changes are submitted to the change control board for review.

**Project Management Plan Updates** The following elements of the project management plan may need to be updated as a result of this process:

- Procurement management plan
- Cost, scope, and schedule baselines

**Project Document Updates** The following documents may need to be updated as a result of this process:

- Requirements documentation and requirements traceability matrix
- Risk register

### Exam Essentials

**Be able to describe the purpose of the Acquire Project Team process.** The Acquire Project Team process involves acquiring and assigning human resources, both internal and external, to the project.

**Be able to describe the purpose of the Conduct Procurements process.** Conduct Procurements involves obtaining bids and proposals from vendors, selecting a vendor, and awarding a contract.

**Be able to name the tools and techniques of the Conduct Procurements process.** The tools and techniques of the Conduct Procurements process are bidder conferences, independent estimates, procurement negotiation, advertising, Internet search, expert judgment, and proposal evaluation techniques.

**Be able to name the contracting life cycle stages.** The stages of the contracting life cycle are requirement, requisition, solicitation, and award.

# Maximizing Team Performance

Maximizing team performance involves leading, mentoring, training, and motivating team members. The project manager leads these efforts with the guidance of the documented human resource plan. The processes used to achieve maximum team performance include Develop Project Team and Manage Project Team, which are both a part of the Project Human Resource Management Knowledge Area.

To be successful, project managers will need to be familiar with several leadership theories, the stages of team development, motivational theories, and leadership styles. Being familiar with and understanding conflict resolution techniques and types of power are also important when managing teams.

## Develop Project Team

The Develop Project Team process is concerned with creating a positive environment for team members; developing the team into an effective, functioning, coordinated group; and increasing the team's competency levels. The proper development of the team is critical to a successful project. Since teams are made up of individuals, individual development becomes a critical factor to project success.

Figure 4.3 shows the inputs, tools and techniques, and outputs of the Develop Project Team process.

NOTE

For more detailed information on the Develop Project Team process, see Chapter 8 of *PMP: Project Management Professional Exam Study Guide, 6th Edition.*

**FIGURE 4.3** Develop Project Team process

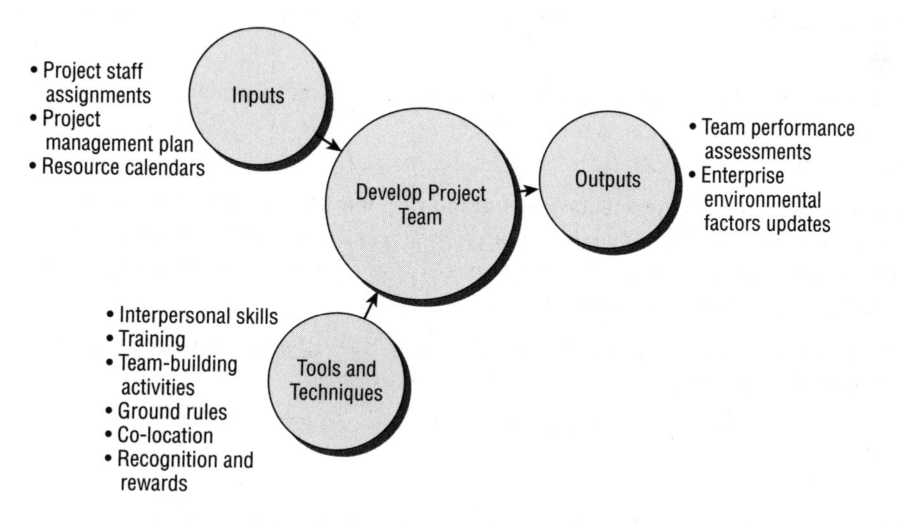

## Inputs of Develop Project Team

The Develop Project Team process includes the following inputs:

**Project Staff Assignments** The list of project team members is obtained through the project staff assignments created from within the Acquire Project Team process.

**Project Management Plan** The human resource plan, a subsidiary plan of the project management plan, is utilized in this process.

**Resource Calendars** Resource calendars provide details about the availability of team members. This is important when coordinating team development activities to ensure participation from all team members.

## Tools and Techniques of Develop Project Team

Before we review the tools and techniques of the Develop Project Team process, it is helpful to understand some key concepts and theories about leadership, management, and motivation.

### Leadership vs. Management

Project managers need to use the traits of both leaders and managers at different times during a project. Therefore, it's important to understand the difference between the two.

Leaders display the following characteristics:

- Impart vision, motivate and inspire, and are concerned with strategic vision
- Have a knack for getting others to do what needs to be done
- Use and understand the following two techniques:
  - Power, which is the ability to get people to do what they wouldn't do ordinarily and the ability to influence behavior
  - Politics, which imparts pressure to conform regardless of whether people agree with the decision
- Have committed team members who believe in their vision
- Set direction and time frames and have the ability to attract good talent to work for them
- Are directive in their approach but allow for plenty of feedback and input
- Have strong interpersonal skills and are well respected

Managers display the following characteristics:

- Are generally task-oriented and concerned with issues such as plans, controls, budgets, policies, and procedures
- Are generalists with a broad base of planning and organizational skills

- Have a primary goal of satisfying stakeholder needs
- Possess motivational skills and the ability to recognize and reward behavior

There are four particularly notable theories regarding leadership and management:

**Theory X & Y** Douglas McGregor defined two models of worker behavior, Theory X and Theory Y, that attempt to explain how different managers deal with their team members. Theory X managers believe the following statements to be true:

- Most people do not like work and will try to steer clear of it.
- People have little to no ambition and need constant supervision.
- People won't actually perform the duties of their job unless threatened.

As a result, Theory X managers are like dictators and impose very rigid controls over their employees. They believe people are motivated only by punishment, money, or position.

Theory Y managers have the following characteristics:

- Believe people are interested in performing their best given the right motivation and proper expectations
- Provide support to their teams
- Are concerned about their team members
- Are good listeners

Theory Y managers believe people are creative and committed to the project goals, that they like responsibility and seek it out, and that they are able to perform the functions of their positions with limited supervision.

**Theory Z** Theory Z was developed by William Ouchi. This theory is concerned with increasing employee loyalty to their organization and results in increased productivity. Theory Z has the following characteristics:

- Puts an emphasis on the well-being of the employees both at work and outside of work
- Encourages steady employment
- Leads to high employee satisfaction and morale

Theory Z develops employee loyalty through group decision making, which also results in a sense of being valued and respected.

Figure 4.4 highlights the differences between a Theory X manager, a Theory Y manager, and a Theory Z manager.

**FIGURE 4.4** Theory X & Y and Theory Z

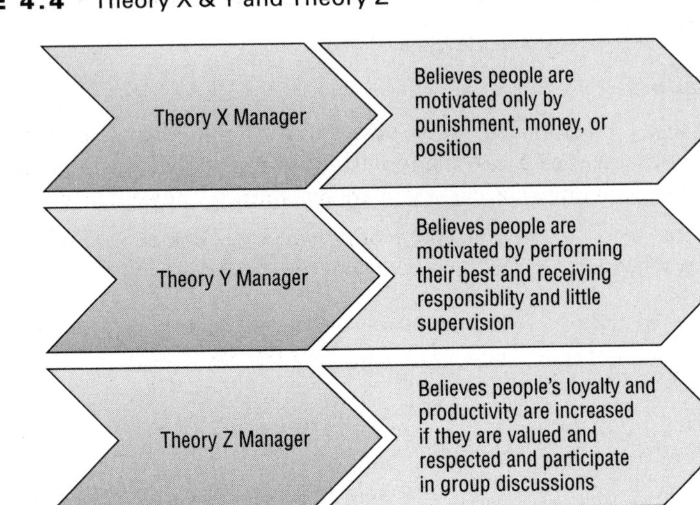


**Contingency Theory** The Contingency theory builds on a combination of Theory Y behaviors and the Hygiene theory. (The Hygiene theory is discussed shortly, in the section "Motivational Theories.") The Contingency theory says that people are motivated to achieve levels of competency and will continue to be motivated by this need even after competency is reached.

**Situational Theory** Paul Hersey and Ken Blanchard developed the Situational Leadership theory during the mid-1970s. This theory's main premise is that the leadership style you use depends on the situation. Both Hersey and Blanchard went on to develop their own situational leadership models. Blanchard's model, Situational Leadership II, describes four styles of leadership that depend on the situation: directing, coaching, supporting, and delegating.

## Leadership Styles

There are various leadership styles that a project manager may use, and in many cases, the appropriate leadership style is based on the situation. The following are examples of leadership styles:

**Autocratic** All decisions are made by the leader.

**Laissez-fair** The leader uses a hands-off approach and allows the team to drive the decisions.

**Democratic** Leaders gather all facts and receive input from the team before reaching a decision.

**Situational** Based on the Blanchard theory of situational leadership that uses four styles: directing, coaching, supporting, and delegating.

**Transactional and Transformational** Developed by Bernard Bass. He described transactional leaders as activity-focused and autonomous; they use contingent reward systems and manage by exception. Transformational leaders are described as focusing on relationships rather than activities.

### The Power of Leaders

Leaders, managers, and project managers use power to convince others to do tasks a specific way. The kind of power they use to accomplish this depends on their personality, their personal values, and the company culture. Here are five forms of power leaders may use:

- Punishment power, also known as *coercive* or *penalty power*, is exerted when employees are threatened with consequences if expectations are not met.

Punishment power should be used as a last resort and only after all other forms have been exhausted.

- Expert power is exerted when the person being influenced believes the manager, or the person doing the influencing, is knowledgeable about the subject or has special abilities that make them an expert.
- Legitimate power, also known as *formal power*, is exerted when power comes about as a result of the influencer's position.
- Referent power is obtained by the influencer through a higher authority.
- Reward power is the ability to grant bonuses and incentives for a job well done.

### Motivational Theories

There are many theories on motivation, and as a project manager, you should understand and tailor the recognition and rewards programs around the project team. The following are some important motivational theories to note:

**Maslow's Hierarchy of Needs** Abraham Maslow theorized that humans have five basic sets of needs arranged in hierarchical order. The idea is that each set of needs must be met before a person can move to the next level of needs in the hierarchy. Once that need is met, they progress to the next level, and so on. Maslow's Hierarchy of Needs theory suggests that once a lower-level need has been met, it no longer serves as a motivator, and the next-higher level becomes the driving motivator in a person's life. Maslow theorized that humans are always in one state of need or another and they can move up and down the pyramid throughout their lives. The following is a brief review of the needs, as shown in Figure 4.5, starting with the highest level and ending with the lowest:

- Self-actualization—Highest level; performing at your peak potential
- Self-esteem needs—Accomplishment, respect for self, capability
- Social needs—A sense of belonging, love, acceptance, friendship
- Safety and security needs—Your physical welfare and the security of your belongings
- Basic physical needs—Lowest level; food, clothing, shelter

**FIGURE 4.5** Maslow's hierarchy of needs

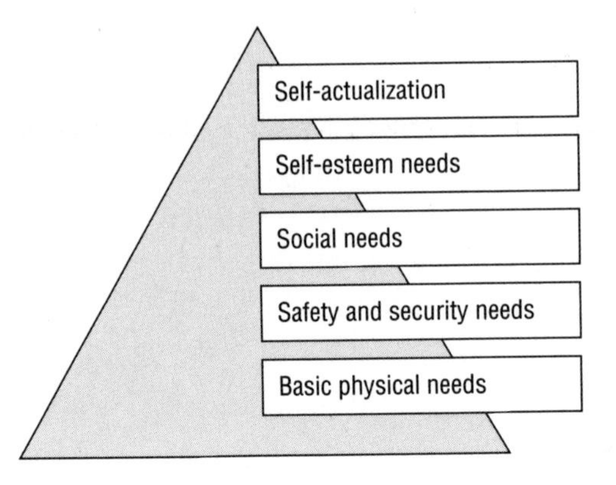


**Hygiene Theory** Frederick Herzberg came up with the Hygiene theory, also known as the Motivation-Hygiene theory. He postulated that two factors contribute to motivation:

- Hygiene factors, which prevent dissatisfaction, deal with work environment issues and include factors such as pay, benefits, conditions of the work environment, and relationships with peers and managers.
- Motivators, which lead to satisfaction, deal with the substance of the work itself and the satisfaction one derives from performing the functions of the job. According to Herzberg, the ability to advance, the opportunity to learn new skills, and the challenges involved in the work are all motivators.

**Expectancy Theory** The Expectancy theory, first proposed by Victor Vroom, says that the expectation of a positive outcome drives motivation. People will behave in certain ways if they think there will be good rewards for doing so and if they themselves value the reward. Also note that this theory says the strength of the expectancy drives the behavior. This means the expectation or likelihood of the reward is linked to the behavior. This theory also says that people become what you expect of them.

**Achievement Theory** The Achievement theory, attributed to David McClelland, says that people are motivated by the need for three things:

- Achievement, as a result of a need to achieve or succeed
- Power, which involves a desire to influence the behavior of others
- Affiliation, as a result of a need to develop relationships with others

## Tools and Techniques

You should know the following tools and techniques of the Develop Project Team process:

- Interpersonal skills
- Training
- Team-building activities

- Ground rules
- Co-location
- Recognition and rewards

**Interpersonal Skills** Interpersonal skills, often referred to as soft skills, include things such as these:

- Leadership
- Influence
- Negotiation
- Communications
- Empathy
- Creativity

By possessing these skills, the project management team can reduce issues within the project team and better manage the team overall.

**Training** Training is a matter of assessing the team members' skills and abilities, assessing the project needs, and providing the training necessary for the team members to carry out their assigned activities. Training can sometimes be a reward as well. Training needs may be incorporated into the staffing management plan or scheduled as they are assessed.

**Team-Building Activities** Part of the project manager's job is to bring the team together, get its members headed in the right direction, and provide motivation, reward, and recognition. This is done using a variety of team-building techniques and exercises. Team building involves getting a diverse group of people to work together in the most efficient and effective manner possible.

Important to team-building activities are the theories behind team development. Authors Bruce Tuckman and Mary Ann Jensen developed a model that describes how teams develop and mature. According to Tuckman and Jensen, all newly formed teams go through the following five stages of development:

**Forming** Forming occurs at the beginning stage of team formation, when team members are brought together, introduced, and told the objectives of the project. Team members tend to behave in a formal and reserved manner.

**Storming** Storming is the stage when action begins. Team members tend to be confrontational with one another as they're vying for position and control.

**Norming** Norming is the stage when team members are comfortable with one another and begin to confront project concerns and issues.

**Performing** Performing is the stage when the team becomes productive and effective. Trust among team members is high, and this is considered to be the mature development stage.

**Adjourning** The adjourning stage is when the team is released after the work is completed.

Figure 4.6 shows the stages of a team's development.

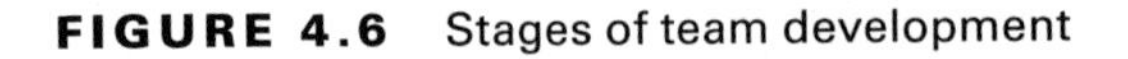

**FIGURE 4.6** Stages of team development

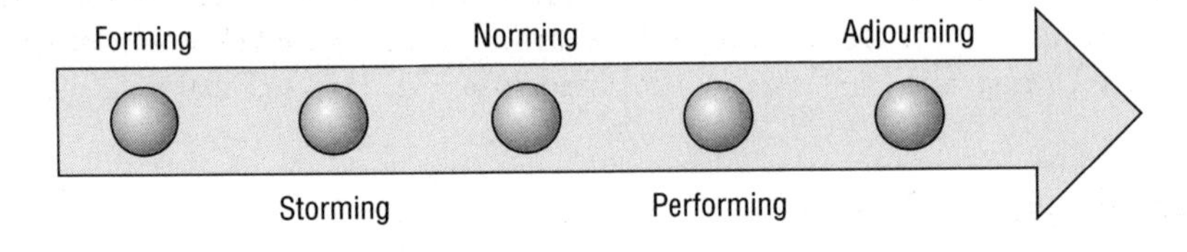


**Ground Rules** Ground rules are expectations set by the project manager and project team that describe acceptable team behavior. Outlining ground rules helps the team understand expectations regarding acceptable behavior and increases productivity.

**Co-location** Co-location refers to basing team members out of the same physical location. Co-location enables teams to function more effectively than if they're spread out among different localities. One way to achieve co-location might be to set aside a common meeting room, sometimes called a war room, for team members who are located in different buildings or across town to meet and exchange information.

**Recognition and Rewards** Recognition and rewards are an important part of team motivation. They are formal ways of recognizing and promoting desirable behavior and are most effective when carried out by the management team and the project manager. Criteria for rewards should be developed and documented, and rewards should be given to team members who go above and beyond the call of duty.

Motivation can be extrinsic or intrinsic:

- Extrinsic motivators are material rewards and might include bonuses, the use of a company car, stock options, gift certificates, training opportunities, extra time off, and so on.
- Intrinsic motivators are specific to the individual. Cultural and religious influences are forms of intrinsic motivators as well.

The recognition and rewards tool of this process is an example of an extrinsic motivator.

## Outputs of Develop Project Team

The following two outputs result from the Develop Project Team process:

**Team Performance Assessments** Team performance assessments involve determining the project team's effectiveness. Assessing these characteristics helps to determine where (or whether) the project team needs improvements. The following indicators are among those assessed:

- Improvements in the team's skills
- Improvements in the team's competencies

- Lower staff attrition
- Greater level of team cohesiveness

**Enterprise Environmental Factors Updates** Updates to the enterprise environmental factors typically include updates to team records regarding training and skill assessment and any personnel administration updates that resulted from carrying out this process.

## Manage Project Team

The Manage Project Team process is concerned with tracking and reporting on the performance of individual team members. During this process, performance appraisals are prepared and conducted, issues are identified and resolved, and feedback is given to the team members. This process involves management skills that promote teamwork and result in high-performance teams.

Figure 4.7 shows the inputs, tools and techniques, and outputs of the Manage Project Team process.

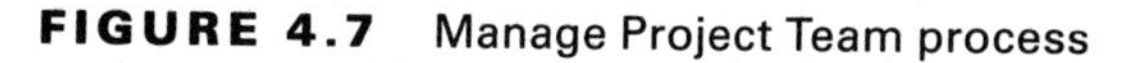

**FIGURE 4.7** Manage Project Team process

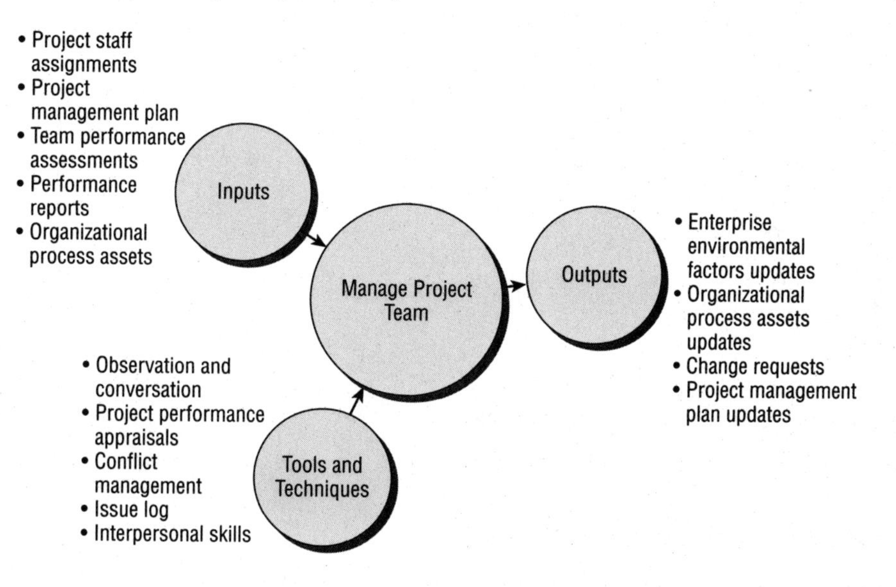

For more detailed information on the Manage Project Team process, see Chapter 8 of *PMP: Project Management Professional Exam Study Guide, 6th Edition*.

## Inputs of Manage Project Team

There are five inputs of the Manage Project Team process:

- Project staff assignments
- Project management plan
- Team performance assessments
- Performance reports
- Organizational process assets

**Project Staff Assignments** Project staff assignments provide the list of project team members and their assignments for use in this process.

**Project Management Plan** Within the project management plan, the human resource plan provides the following information that is utilized in this process:

  - Staffing management plan
  - Roles and responsibilities
  - The project organization chart

**Team Performance Assessments** Team performance assessments, which are an output of the Develop Project Team process, provide necessary information and insight into the performance and issues of the project team. This information is necessary for managing the group and taking action to resolve any issues and to foster improvement.

**Performance Reports** Performance reports document the status of the project compared to the forecasts. For additional information on performance reports, see the outputs of the Report Performance process in Chapter 5, "Monitoring and Controlling the Project," in this book.

**Organizational Process Assets** Organizational process assets utilized within this process include any existing organizational perks and forms of recognition.

## Tools and Techniques of Manage Project Team

Know the following tools and techniques of the Manage Project Team process:

- Observation and conversation
- Project performance appraisals
- Conflict management
- Issue log
- Interpersonal skills

**Observation and Conversation** Observing team performance can be used as an assessment tool. Both observation and conversation help the project management team stay tuned in to the attitudes and feelings of the project team members, making both an important tool for communication.

**Project Performance Appraisals** Project performance appraisals are typically annual or semiannual affairs where managers let their employees know what they think of their performance and rate them accordingly. These are usually manager-to-employee exchanges but can incorporate a 360-degree review, which takes in feedback from nearly everyone the team member interacts with, including key stakeholders, customers, project manager, peers, subordinates, and the delivery person if they have a significant amount of project interaction.

Performance appraisal time is also a good time to do the following:

- Assess training needs
- Clarify roles and responsibilities
- Set goals for the future

**Conflict Management** Conflict exists when the desires, needs, or goals of one party are incompatible with the desires, needs, or goals of another party (or parties). Conflict is the incompatibility of goals, which often leads to one party resisting or blocking the other party from attaining their goals. Conflict management is how an individual deals with these types of scenarios or issues.

The following six methods of conflict management are also shown in Figure 4.8:

**Forcing** One person forces a solution on the other parties. This is an example of a win-lose conflict resolution technique.

**FIGURE 4.8** Conflict resolution techniques

| Conflict Resolution Techniques | | | | | |
|---|---|---|---|---|---|
| Forcing (win-lose) | Smoothing/ Accomodating (lose-lose) | Compromise (lose-lose) | Confrontation/ Problem Solving (win-win, BEST option) | Collaborating (win-win) | Withdrawal/ Avoidance (lose-lose, WORST option) |

**Smoothing/Accommodating** Smoothing/accommodating is a temporary way to resolve conflict in which the areas of agreement are emphasized over the areas of difference, so the real issue stays buried. This is an example of a lose-lose conflict resolution technique because neither side wins.

**Compromise** Compromise is achieved when each of the parties involved in the conflict gives up something to reach a solution. This is an example of a lose-lose conflict resolution technique because neither side gets what they wanted.

**Confrontation/Problem Solving** Confrontation and problem solving are the best ways to resolve conflict. One of the key actions performed with this technique is a fact-finding mission. This is the conflict resolution approach project mangers use most often and is an example of a win-win conflict resolution technique.

**Collaborating** When multiple viewpoints are discussed and shared and team members have the opportunity to examine all the perspectives of the issue, collaboration occurs. Collaborating will lead to true consensus with team members committing to the decision.

**Withdrawal/Avoidance** An example of withdrawal/avoidance is when one of the parties gets up, leaves, and refuses to discuss the conflict. This is an example of a lose-lose conflict resolution technique because no solution is ever reached. This is considered to be the worst conflict resolution technique.

The following should also be noted about conflict and conflict resolution:

- In any situation, conflict should be dealt with as soon as it arises.
- According to the *PMBOK® Guide*, when you have successfully resolved conflict, it will result in increased productivity and better, more positive working relationships.
- Most conflicts are a result of schedule issues, availability of resources (usually the lack of availability), or personal work habits.
- Team members should be encouraged to resolve their own conflicts.
- Ground rules, established policies, and procedures help mitigate conflict before it arises.

**Issue Log** The issue log is a place to document the issues that prevent the project team from meeting project goals. In addition to the issue, the following should be noted:

- Who is responsible for resolving the issue
- Date the resolution is needed

**Interpersonal Skills** There are three types of interpersonal skills used most often in this process:

- Leadership
- Influence
- Effective decision making

## Outputs of Manage Project Team

The Manage Project Team process includes the following outputs:

- Enterprise environmental factors updates
- Organizational process assets updates
- Change requests
- Project management plan updates

**Enterprise Environmental Factors Updates** There are two components of the enterprise environmental factors that may need updating as a result of this process:

- Input to organizational performance appraisals from team members who have a high level of interaction with the project and one another
- Personnel skill updates

**Organizational Process Assets Updates** The organizational process asset updates output has three components:

- Historical information and lessons learned documentation
- Templates
- Organizational standard processes

**Change Requests** Change requests may result from a change in staffing. Also included within change requests are corrective actions that may result from disciplinary actions or training needs and preventive actions, which may be needed to reduce potential issues among the project team.

**Project Management Plan Updates** Project management plan updates may include changes that occurred to the staffing management plan or the human resource plan.

## Exam Essentials

**Be able to describe the purpose of the Develop Project Team process.** The Develop Project Team process is concerned with creating a positive environment for team members; developing the team into an effective, functioning, coordinated group; and increasing the team's competency levels.

**Be able to name the five stages of group formation.** The five stages of group formation are forming, storming, norming, performing, and adjourning.

**Be able to define Maslow's highest level of motivation.** Self-actualization occurs when a person performs at their peak and all lower-level needs have been met.

**Be able to name the five types of power.** The five levels of power are reward, punishment, expert, legitimate, and referent.

**Be able to identify the six styles of conflict resolution.** The six styles of conflict resolution are forcing, smoothing, compromising, confrontation, collaborating, and withdrawal.

**Be able to name the tools and techniques of the Manage Project Team process.** The tools and techniques of Manage Project Team are observation and conversation, project performance appraisal, conflict management, issue log, and interpersonal skills.

# Executing the Project Management Plan

Executing the project management plan is carried out through the Direct and Manage Project Execution process, which is a process belonging to the Project Integration Management Knowledge Area. This is where the project work is carried out, according to the plan, as a way of producing the deliverables on time and on budget. The Direct and Manage Project Execution process is also responsible for implementing changes that have been approved by the change control board. Typically, approved changes are implemented according to the change management plan as a way of meeting project requirements.

When implementing the plan, the project manager will need to be proactive in managing project communications. This includes distributing relevant project information in an efficient and effective manner. Managing stakeholder expectations is also important at this stage. Both of these objectives can be accomplished by carrying out the Distribute Information and Management Stakeholder Expectations processes in accordance with the communications management plan.

## Direct and Manage Project Execution

The Direct and Manage Project Execution process is responsible for carrying out the project management plan. The project manager oversees the actual work, staying on top of issues and problems and keeping the work lined up with the project plan. Coordinating and integrating all the elements of the project are among the most challenging aspects of this process.

According to the *PMBOK® Guide*, this process also requires implementing corrective actions to bring the work of the project back into alignment with the project plan, preventive actions to reduce the probability of negative consequences, and repairs to correct product defects discovered during the quality processes.

Figure 4.9 shows the inputs, tools and techniques, and outputs of the Direct and Manage Project Execution process.

**FIGURE 4.9** Direct and Manage Project Execution process

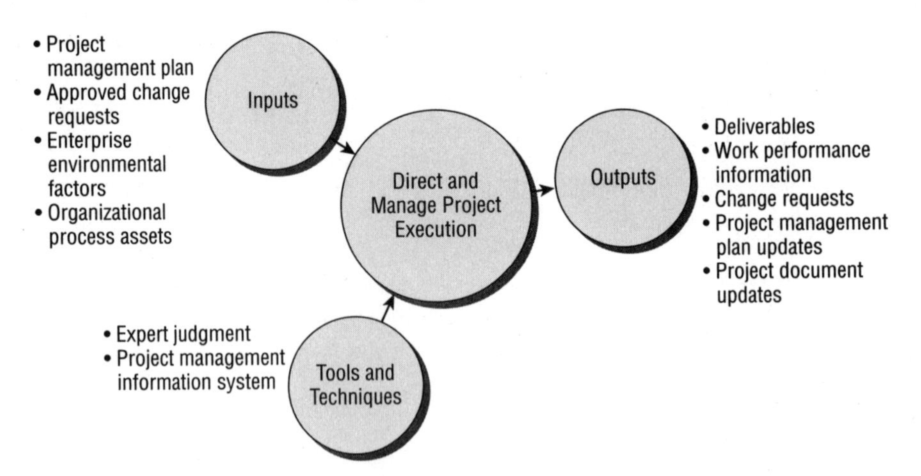

For more detailed information on the Direct and Manage Project Execution process, see Chapter 8 in *PMP: Project Management Professional Exam Study Guide, 6th Edition.*

## Inputs of Direct and Manage Project Execution

Know the following inputs of the Direct and Manage Project Execution process:

- Project management plan
- Approved change requests
- Enterprise environmental factors
- Organizational process assets

**Project Management Plan** The project management plan documents the collection of outputs of the planning processes and describes and defines the work to be carried out and how the project should be executed, monitored, controlled, and closed.

**Approved Change Requests** Approved change requests come about as a result of the change request status updates output of the Perform Integrated Change Control process. Approved change requests are then submitted as inputs to the Direct and Manage Project Execution process for implementation. Implementation of approved changes may also include the implementation of workarounds, which are unplanned responses to negative risks that have occurred.

**Enterprise Environmental Factors** The following enterprise environmental factors are considered when performing the Direct and Manage Project Execution process:

- Company culture and organizational structure
- Facilities available to the project team
- Personnel guidelines and hiring practices
- Risk tolerance levels
- Project management information systems

**Organizational Process Assets** The following organizational process assets are utilized within the Direct and Manage Project Execution process:

- Historical information from past projects
- Organizational guidelines and work processes
- Process measurement databases
- Issue and defect databases

## Tools and Techniques of Direct and Manage Project Execution

You should be familiar with the following tools and techniques of the Direct and Manage Project Execution process:

**Expert Judgment** Expert judgment is provided by the project manager and project team members as well as by the following sources:

- Industry and professional associations
- Consultants
- Stakeholders
- Other units within the company

**Project Management Information System** As used in the Direct and Manage Project Execution process, the project management information system (PMIS) provides the ability to connect to automated tools, such as scheduling software and configuration management systems, that can be utilized during project execution.

## Outputs of Direct and Manage Project Execution

These five outputs result from carrying out the Direct and Manage Project Execution process:

- Deliverables
- Work performance information
- Change requests
- Project management plan updates
- Project document updates

**Deliverables** According to the *PMBOK® Guide*, a deliverable (see Figure 4.10) is defined as any unique and verifiable product, result, or capability to perform a service that must be produced to complete a process, phase, or project.

**FIGURE 4.10** Definition of a deliverable

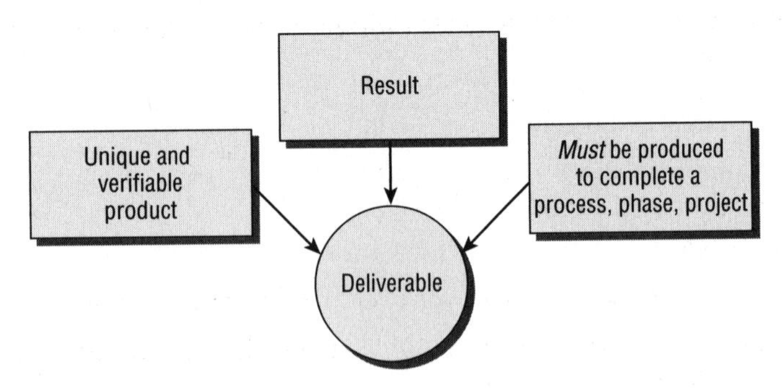

During the Direct and Manage Project Execution process, the following information regarding the outcomes of the work is gathered and recorded:

- Activity completion dates
- Milestone completions
- Status of the deliverables
- Quality of the deliverables
- Costs
- Schedule progress and updates

Although the *PMBOK® Guide* breaks up these processes for ease of explanation, several of the Executing and Monitoring and Controlling processes are performed together. Remember that the processes are iterative.

**Work Performance Information** Work performance information involves gathering, documenting, and recording the status of project work. Here are types of information that may be gathered during this process:

- Schedule status and progress
- Status of deliverable completion
- Progress and status of schedule activities
- Adherence to quality standards
- Status of costs (those authorized and costs incurred to date)
- Schedule activity completion estimates for those activities started
- Percentage of schedule activities completed
- Lessons learned
- Resource consumption and utilization

**Change Requests** Changes can come about from several sources, including stakeholder requests, external sources, and technological advances. Change requests may encompass changes to the following:

- Schedule
- Scope
- Requirement
- Resource changes

Implementation of change requests may incorporate the following actions:

**Corrective Actions** Corrective actions are taken to get the anticipated future project outcomes to align with the project plan.

**Preventive Actions** Preventive actions involve anything that will reduce the potential impacts of negative risk events should they occur. Contingency plans and risk responses are examples of preventive action.

**Defect Repairs** Defects are project components that do not meet the requirements or specifications. Defect repairs are discovered during quality audits or when inspections are performed. A *validated defect repair* is the result of a reinspection of the original defect repair.

**Project Management Plan Updates** The following project management plan updates may occur as a result of the Direct and Manage Project Execution process:

- Subsidiary project plans
- Project baselines

**Project Document Updates** The following project documents may undergo updates as a result of the Direct and Manage Project Execution process:

- Requirements documentation and project logs
- Risk register
- Stakeholder register

## Distribute Information

The Distribute Information process is responsible for getting information about the project to stakeholders in a timely manner. It describes how reports, and other information, are distributed and to whom. Executing the communications management plan also occurs during this process.

Figure 4.11 shows the inputs, tools and techniques, and outputs of the Distribute Information process.

**FIGURE 4.11** Distribute Information process

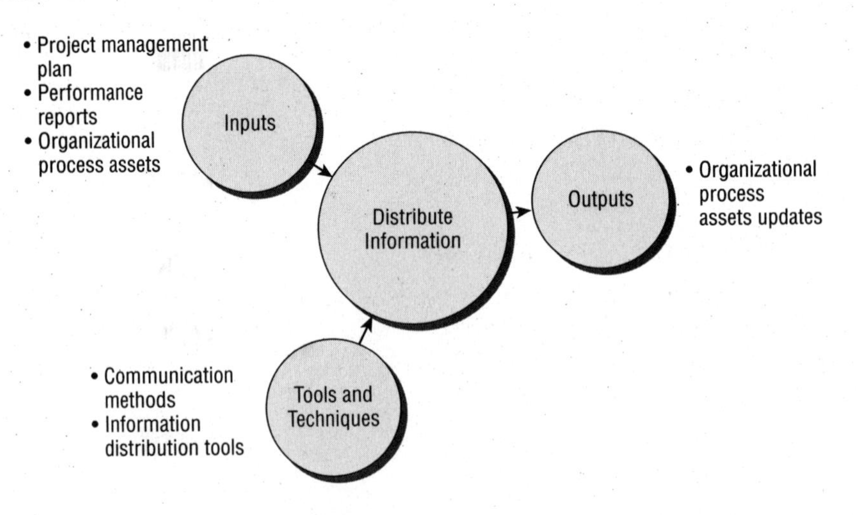

For more detailed information on the Distribute Information process, see Chapter 9 of *PMP: Project Management Professional Exam Study Guide, 6th Edition.*

## Inputs of Distribute Information

Know the following inputs of the Distribute Information process:

**Project Management Plan** The communications management plan, a subsidiary plan of the project management plan, is put into action during this process.

**Performance Reports** Performance reports are utilized to distribute performance and status information to the appropriate individuals.

**Organizational Process Assets** For this process, organizational process assets refer to any policies, procedures, and guidelines that the organization already has in place for distributing information and any existing documents and information that are relevant to the process.

## Tools and Techniques of Distribute Information

The tools and techniques of the Distribute Information process include the following two items:

**Communication Methods** Communication methods include all means feasible to communicate project information to the appropriate individuals:

- Meetings
- Email messages
- Videoconferences
- Conference calls
- Other remote access tools or methods

**Information Distribution Tools** The following information distribution tools are used to get the project information to the project team or stakeholders:

- Electronic communication
- Electronic project management tools
- Hard copy

## Outputs of Distribute Information

There is only one output of the Distribute Information process that you should know: updates to the organizational process assets.

There can be several updates to the organizational process assets as a result of carrying out this process. These updates are listed and described in Table 4.1.

**TABLE 4.1** Organizational process assets updates

| Update Item | Description |
|---|---|
| Stakeholder notifications | Notifications sent to stakeholders when solutions and approved changes have been implemented, the project status has been updated, issues have been resolved, and so on. |
| Project reports | The project status reports and minutes from project meetings, lessons learned, closure reports, and other documents from all the process outputs throughout the project. |
| Project presentations | Project information presented to the stakeholders and other appropriate parties when necessary. |
| Project records | Memos, correspondence, and other documents concerning the project. |
| Feedback from stakeholders | Feedback that can improve future performance on this project or future projects. The feedback should be captured and documented. |
| Lessons learned documentation | Information that is gathered and documented throughout the course of the project and can be used to benefit the current project, future projects, or other projects being performed by the organization. |

## Manage Stakeholder Expectations

The Manage Stakeholder Expectations process is responsible for satisfying the needs of the stakeholders by managing communications with them, resolving issues, improving project performance by implementing requested changes, and managing concerns in anticipation of potential problems.

Figure 4.12 shows the inputs, tools and techniques, and outputs of the Manage Stakeholder Expectations process.

For more detailed information on the Manage Stakeholder Expectations process, see Chapter 9 of *PMP: Project Management Professional Exam Study Guide, 6th Edition.*

**FIGURE 4.12** Manage Stakeholder Expectations process

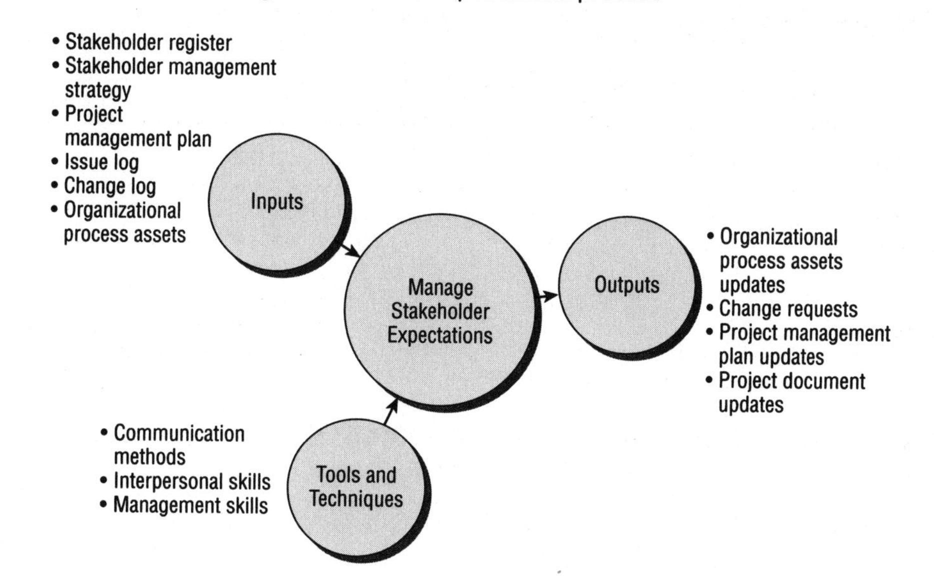


## Inputs of Manage Stakeholder Expectations

Know the following inputs of the Manage Stakeholder Expectations process:

- Stakeholder register
- Stakeholder management strategy
- Project management plan
- Issue log
- Change log
- Organizational process assets

**Stakeholder Register** To manage stakeholder expectations, you need to know who the stakeholders are on the project. The stakeholder register provides this information.

**Stakeholder Management Strategy** The stakeholder management strategy document contains a tailored strategy assembled for managing stakeholder expectations within the project.

**Project Management Plan** The communications management plan, which documents stakeholder goals, objectives, and levels of required communication, is utilized from within the project management plan.

**Issue Log** The issue log, or action log, is used to track and document resolution of any existing stakeholder concerns or issues. Issues are ranked according to their urgency and potential impact on the project and are assigned a responsible party and due date for resolution.

**Change Log** The change log, in relation to this process, keeps stakeholders updated on changes and their impacts to the project.

**Organizational Process Assets** The following organizational process assets influence this process:

- Issue management and change control procedures
- Organizational communication requirements
- Information from previous projects

According to the *PMBOK® Guide,* it's the project manager's responsibility to manage stakeholder expectations.

## Tools and Techniques of Manage Stakeholder Expectations

Be familiar with the following three tools and techniques of the Manage Stakeholder Expectations process, and how they are used in this process:

**Communication Methods** The communication methods technique includes specific strategies that were documented within the communications management plan given to the unique stakeholders identified for this project and their communication needs.

**Interpersonal Skills** Interpersonal skills were introduced with the Develop Project Team process. To manage stakeholder expectations, the project manager utilizes soft skills, such as building trust, establishing relationships, and listening.

**Management Skills** The project manager utilizes management skills such as negotiation, presentation, speaking, and writing to manage stakeholder expectations and bring about the appropriate outcomes and results.

## Outputs of Manage Stakeholder Expectations

The Manage Stakeholder Expectations process results in the following outputs:

- Organizational process assets updates
- Change requests
- Project management plan updates
- Project document updates

**Organizational Process Assets Updates** The organizational process assets may include the following updates:

- Causes of issues
- Explanations of certain corrective actions
- Lessons learned

**Change Requests** The following change requests may emerge as a result of this process:

- Changes to the product or project
- Corrective actions
- Preventive actions

**Project Management Plan Updates** Updates to the project management plan, as a result of carrying out this process, often involve updating the communications management plan.

**Project Document Updates** The following project documents may need an update as a result of carrying out this process:

- Stakeholder management strategy
- Stakeholder register
- Issue log

**Exam Essentials**

**Be able to identify the distinguishing characteristics of Direct and Manage Project Execution.** Direct and Manage Project Execution is where the work of the project is performed, and the majority of the project budget is spent during this process.

**Be able to describe the purpose of the Distribute Information process.** The purpose of the Distribute Information process is to get information to stakeholders about the project in a timely manner.

**Be able to describe the purpose of the Manage Stakeholder Expectations process.** Manage Stakeholder Expectations concerns satisfying the needs of the stakeholders and successfully meeting the goals of the project by managing communications with stakeholders.

# Implementing Approved Changes

As described in the previous section, the need for changes will emerge as the project work is executed. All change requests created to deal with this need must first be approved by the change control board before they can be implemented. This occurs through the Perform Integrated Change Control process, which is discussed in Chapter 5, "Monitoring and Controlling the Project," and must be managed according to the change management plan.

It's also important to keep in mind that the moment the work enters into the executing stage of the project, monitoring and controlling activities begin and continue concurrently as the work is progressing. Monitoring and controlling activities also identify the need for corrective and preventive actions as well as defect repairs. Once change requests are approved, they are implemented through the Direct and Manage Project Execution process, discussed previously.

**Exam Essentials**

**Be able to describe the various steps that a change request progresses through.** Change requests are reviewed by the change control board through the Perform Integrated Change Control process. If approved, the changes are then implemented through the Direct and Manage Project Execution process.

# Implementing the Quality Management Plan

Implementing the quality management plan occurs throughout the Executing and Monitoring and Controlling process groups. During Executing, quality assurance activities take place. Quality assurance is responsible for auditing quality control activities and the effectiveness of the project management processes. These activities are carried out through the Perform Quality Assurance process.

The Perform Quality Assurance process involves performing systematic quality activities and uses quality audits to determine which processes should be used to achieve the project requirements and to assure that they are performed efficiently and effectively. This process also brings about continuous process improvement through improved process performance and eliminating unnecessary actions. This is done by assessing whether the processes are efficient and whether they can be improved.

Figure 4.13 shows the inputs, tools and techniques, and outputs of the Perform Quality Assurance process.

**FIGURE 4.13** Perform Quality Assurance process

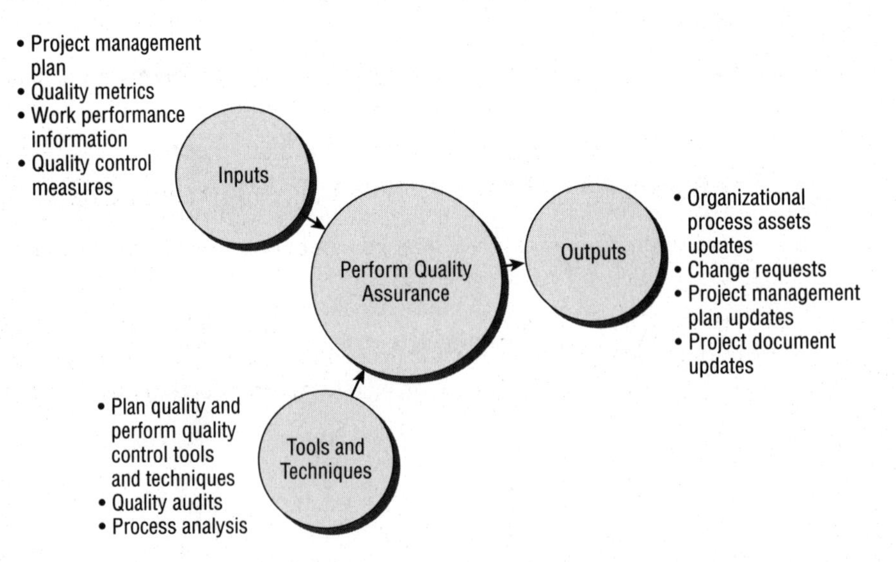

For more detailed information on the Perform Quality Assurance process, see Chapter 9 of *PMP: Project Management Professional Exam Study Guide, 6th Edition.*

## Inputs of Perform Quality Assurance

Know the following inputs of the Perform Quality Assurance process:

- Project management plan
- Quality metrics
- Work performance information
- Quality control measurements

The project team members, project manager, and other stakeholders are all responsible for the quality assurance of the project.

**Project Management Plan** The project management plan provides the following documents that are utilized in this process:

- Quality management plan, which guides the process
- Process improvement plan, which describes the steps for analyzing processes to bring about improvement

**Quality Metrics** Quality metrics provide the necessary information to evaluate the performance and effectiveness of the quality processes.

**Work Performance Information** Results included within the performance reports are utilized during the audit process.

**Quality Control Measurements** Quality control measurements contain results of quality control activities and can be used to evaluate the quality processes.

## Tools and Techniques of Perform Quality Assurance

You should be familiar with the following tools and techniques of the Perform Quality Assurance process:

**Plan Quality and Perform Quality Control Tools and Techniques** The Perform Quality Assurance process utilizes tools and techniques also used during the Plan Quality and Perform Quality Control processes. The following tools and techniques are used in the Plan Quality process:

- Cost-benefit analysis
- Cost of quality

- Control charts
- Benchmarking
- Design of experiments
- Statistical sampling
- Flowcharting
- Proprietary quality management methodologies
- Additional quality planning tools

For a description of the tools and techniques also utilized in the Plan Quality process, see Chapter 3, "Planning the Project," in this book.

The following tools and techniques are used in the Perform Quality Control process:

- Cause-and-effect diagrams
- Control charts
- Flowcharting
- Histogram
- Pareto chart
- Run chart
- Scatter diagram
- Statistical sampling
- Inspection
- Approved change requests review

For a description of the tools and techniques also utilized in the Perform Quality Control process, see Chapter 5 in this book.

**Quality Audits** Quality audits are independent reviews performed by trained auditors or third-party reviewers either on a regular schedule or at random. The purpose of a quality audit is to identify ineffective and inefficient activities or processes used on the project.

Quality improvements come about as a result of the quality audits, as shown in Figure 4.14.

Quality improvements are implemented by submitting change requests, which may entail taking corrective action.

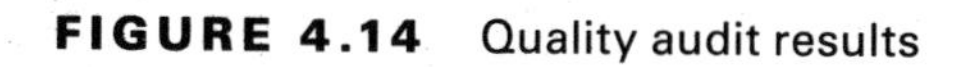

**FIGURE 4.14** Quality audit results

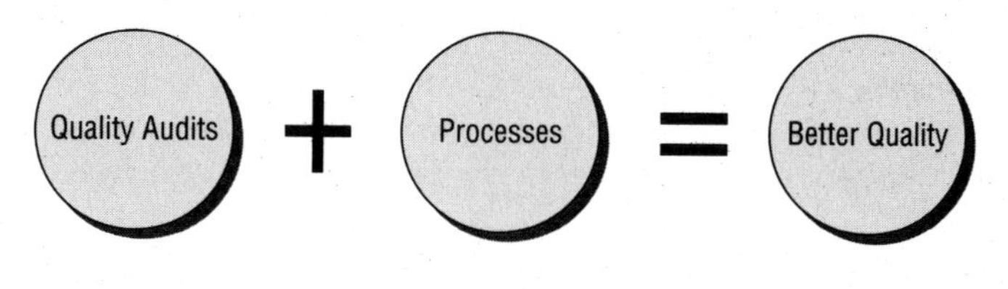


### Benefits of Quality Audits

Performing quality audits results in the following benefits:

- The product of the project is fit for use and meets safety standards.
- Applicable laws and standards are adhered to.
- Corrective action is recommended and implemented where necessary.
- The quality plan for the project is adhered to.
- Quality improvements are identified.
- The implementation of approved change requests, corrective actions, preventive actions, and defect repairs is confirmed.

**Process Analysis** Process analysis looks at process improvement from an organizational and technical perspective. An example of a process analysis technique is root cause analysis.

According to the *PMBOK® Guide*, process analysis follows the steps in the process improvement plan and examines the following:

- Problems experienced while conducting the project
- Constraints experienced while conducting the work of the project
- Inefficient and ineffective processes identified during process operation

## Outputs of Perform Quality Assurance

The Perform Quality Assurance process results in the following four outputs:

- Organizational process assets updates
- Change requests
- Project management plan updates
- Project document updates

**Organizational Process Assets Updates** Within the organizational process assets, the quality standards may require updates as a result of carrying out this process.

**Change Requests** During this process, any recommended corrective actions, whether they are a result of a quality audit or process analysis, should be acted upon immediately. Change requests may involve corrective action, preventive action, or defect repair.

**Project Management Plan Updates** As a result of this process, the quality management plan may need to be updated. The following documents may also require changes within the project management plan:

- Cost management plan
- Schedule management plan

**Project Document Updates** Updates may include changes to the following project documents:

- Quality audit reports
- Process documentation
- Training plans

**Exam Essentials**

**Be able to describe the purpose of the Perform Quality Assurance process.** The Perform Quality Assurance process is concerned with making certain the project will meet and satisfy the quality standards of the project.

# Implementing the Risk Management Plan

As the work of the project is executed, predefined risk triggers may occur and some risks will be realized, calling for the execution of risk response plans, contingency plans, and fallback plans. These responses are defined within the risk register and should be managed according to the risk management plan. The risk management plan is important in reducing the impact of negative risks on the project.

Because the *PMBOK® Guide* addresses the executing activities of risk management within the Monitoring and Controlling process group, this will be further addressed in Chapter 5 in this book.

**Exam Essentials**

**Be able to describe where risk response plans are documented and when they are executed.** Risk response plans, contingency plans, and fallback plans are documented within the risk register and are executed when predefined risk triggers occur.

# Bringing the Processes Together

Let's briefly review the processes that were covered in this chapter and, more importantly, how these processes work together. You may recall that the primary objective of the Executing process group is to complete the work defined in the project management plan. In addition to this, we also coordinate and manage resources, implement approved changes, and distribute information.

As we went through the processes covered in this chapter, you may have noticed an ongoing theme: human resources. Whether these resources were internal to the organization or external, we covered a lot of information that dealt with the project team, including obtaining the resources and managing them. We also covered the management of stakeholder expectations.

Figure 4.15 shows how people-centric the Executing Process Group really is. Within the realms of this figure, we see the following take place:

- The project information is planned.
- The planned information is handed off to the project team for execution of the project work.
- The project team's execution of the work generates results, which must be monitored and controlled.
- The results are fed back into the previous processes as necessary to improve the work and make changes and updates as needed.

**FIGURE 4.15** The project team

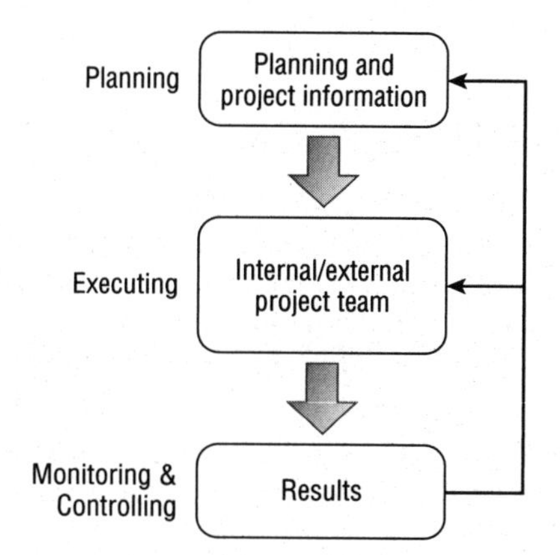


Clearly, a lot of behind-the-scenes effort goes into the executing stage. For example, the project team must be assembled and managed.

Altogether, we covered eight processes within this chapter, which make up the Executing process group. These processes spanned the following Knowledge Areas:

- Project Integration Management
- Project Quality Management

- Project Human Resource Management
- Project Communications Management
- Project Procurement Management

Next, we will review the process interactions that occur within each Knowledge Area during the Executing process group.

## Project Integration Management Knowledge Area Review

During the Executing process group, the Project Integration Management Knowledge Area covers the implementation of the project management plan through a single process known as the Direct and Manage Project Execution process. As Figure 4.16 shows, the project management plan and the approved changes (which occur after the work has already been executed) are carried out through this process. As a result, deliverables and information on the work performance during the project are generated.

**FIGURE 4.16** Process interaction—integration

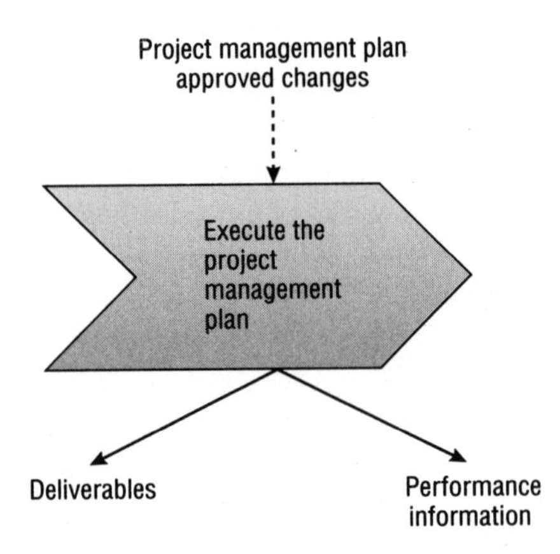


## Project Quality Management Knowledge Area Review

The Perform Quality Assurance process is the only quality-related process that falls within the Executing process group. As you may recall, this is where auditing takes place, which ultimately leads to implemented changes and continuous process improvement. The following items are among those that are audited:

- Quality requirements
- Results from quality control measurements

Figure 4.17 provides a glimpse into the information needed to perform the quality audits. Quality metrics, which were defined in the Planning process group, provide work performance information and quality measurements. These quality measurements are gathered while the project's quality is being monitored and controlled. You can see the level of interaction between the three quality processes throughout the life cycle of the project. The processes are iterative.

**FIGURE 4.17** Process interaction—quality

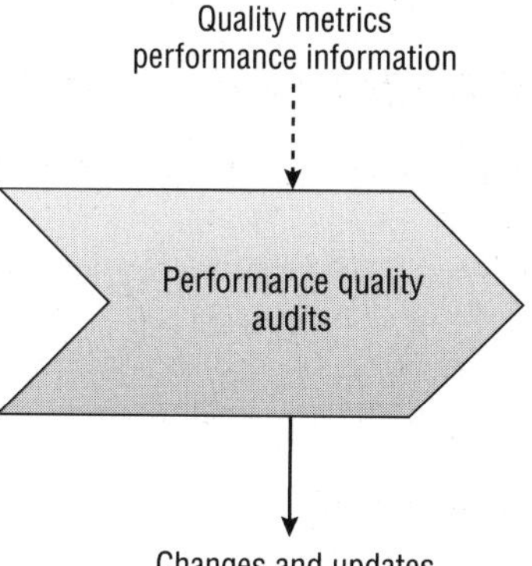


## Project Human Resource Management Knowledge Area Review

Three of the four processes related to human resources occur within the Executing process group. Remember that this process group involves coordinating and managing people, and you cannot implement the project work without a project team. Figure 4.18 shows, step by step, how the project team comes together so that the project work can be rolled out.

**FIGURE 4.18** Process interaction—human resource

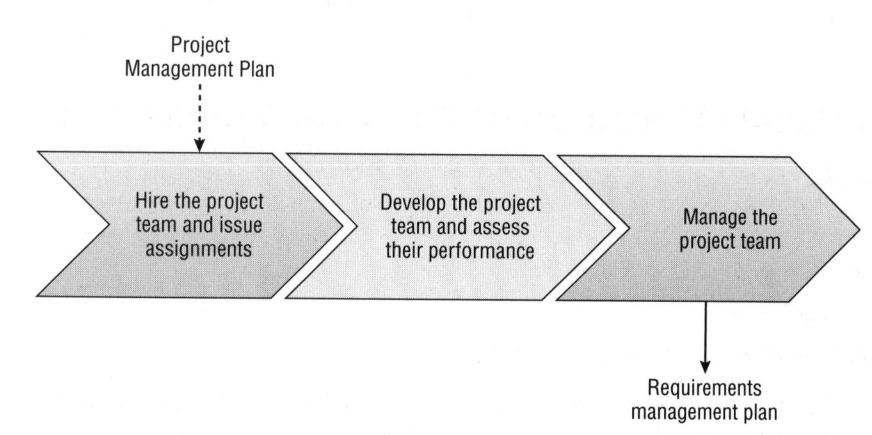

Here is an overview of these steps:

1. Using the project management plan as an information resource, bring the project team on board and issue staff assignments.
2. Throughout project execution, assess the performance of the project team for developmental purposes. This includes enhancing the team's skills, monitoring the level of interaction among the group members, and improving the team's overall performance.
3. Use the performance assessment that was generated in the second step to manage the team. Here, issues are resolved and changes managed. The result? Recommend changes to resolve any staffing issues that may have emerged.

## Project Communications Management Knowledge Area Review

As the project work is executed, communication becomes more important than ever. All of the planning behind how communication will take place within the project goes into effect. This is where knowing the communication needs and requirements of your stakeholders becomes particularly important. Be sure you place a great level of importance on knowing your stakeholders.

There are two communications-related processes within the Executing process group. The first makes project information available to stakeholders. As you can see in Figure 4.19, the stakeholder communication requirements must be developed before you can distribute information. After all, you cannot meet your team's communication needs if you do not know what they are. These communication needs and requirements are determined in the Initiating and Planning process groups.

**FIGURE 4.19** Process interaction—communications

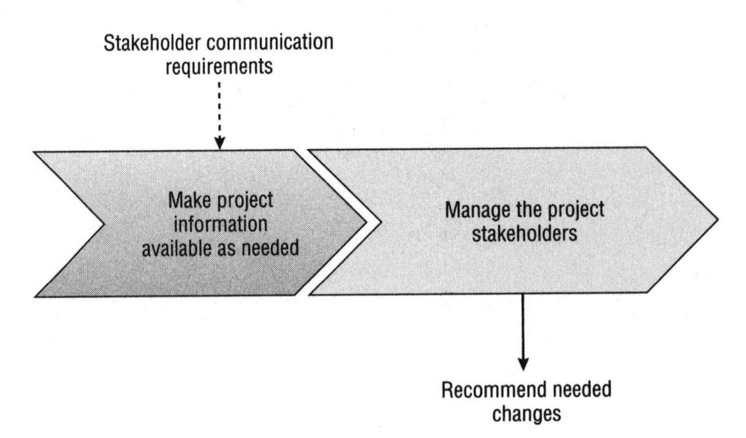

Figure 4.19 also shows how the management of the project stakeholders fits in with the communication processes. This step involves addressing any existing issues and recommending changes to resolve conflicts. It also involves the management of stakeholder expectations with the idea of unifying the project goals.

## Project Procurement Management Knowledge Area Review

During the Executing process group, vendors are selected for work that will be handled externally. Conducting procurements is part of obtaining and hiring the resources needed to complete the project work. As you can see, much of project execution involves hiring and managing people resources—internal and external to the organization.

Figure 4.20 shows how the single procurement process that takes place during project execution results in the selection of the project's sellers and the issuance of procurement contracts. To arrive at this outcome, you will use several planned procurement items:

- Results of the make-or-buy decisions made during the planning phase
- List of qualified vendors (sellers) and their proposals
- Criteria that will be used to select the vendors
- Any agreements or documents impacting vendor selection
- Project management plan, which contains the procurement management plan

**FIGURE 4.20** Process interaction—procurement

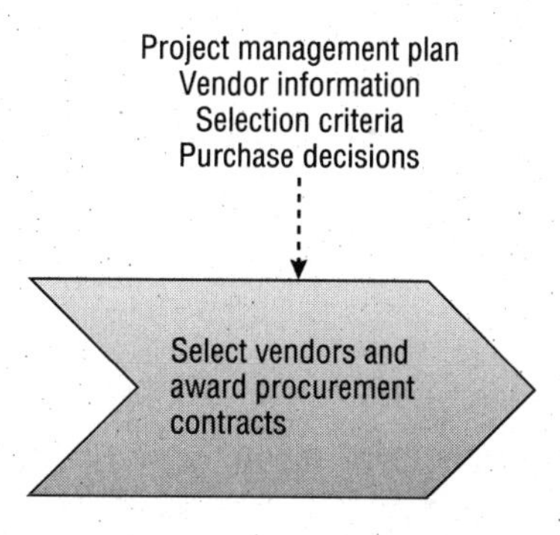


As you reflect on the Executing process group, always remember that the key purpose is to complete the project work.

# Review Questions

1. Jon is gearing up for his upcoming project, which involves programming a complex system that will allow two programs to talk to each other. The project kicks off in one week, and he had been specifically requested by the customer. As a software architect, this will be his most challenging project yet, and he has been looking forward to the assignment for months. Which of the following options BEST describes Jon's upcoming project role?

   **A.** Staff assignment

   **B.** Pre-assignment

   **C.** Project team member

   **D.** Acquisition

2. All of the following are benefits of virtual teams EXCEPT:

   **A.** Access to resources otherwise unavailable

   **B.** Reduction in travel expenses

   **C.** Utilization of a war room

   **D.** Reduction in time spent commuting

3. At which stage of team development do employees compete for control?

   **A.** Forming

   **B.** Storming

   **C.** Norming

   **D.** Performing

4. Sally, a junior project manager for Project Red, is in the process of holding a status meeting for the project manager, who stepped out to deal with a procurement issue. The tension in the room has been high as a result of two critical resources who suddenly quit. John and Rick, both team members, have been arguing for five minutes over who will take over the tasks. Sally has tried to interject multiple times with no success. The side conversations don't make managing the meeting any easier. What is the most likely cause of the inefficient meeting?

   **A.** Absence of the project manager

   **B.** Poorly defined communications management plan

   **C.** Poor recognition and rewards

   **D.** Lack of ground rules

5. Which of the following levels within Maslow's hierarchy of needs describes the need to belong?

   **A.** Social needs

   **B.** Self-esteem needs

   **C.** Self-actualization

   **D.** Safety and security needs

6. Which of the following conflict management techniques is said to be the BEST strategy?
   A. Forcing
   B. Smoothing
   C. Collaborating
   D. Confronting

7. Meetings, email, videoconferences, and conference calls are all examples of which of the following options?
   A. Project management tools
   B. Communication distribution tools
   C. Communication methods
   D. Communication types

8. The project team of a new housing expansion project is in a state of frenzy because of a surprise quality audit that is being conducted today on the project. In three days, a major deliverable is due. The project manager explains to the team that the quality audit is important. What reasoning is the project manager likely to give the team for explaining why the quality audits are needed in a project?
   A. Quality audits look at problems experienced while conducting the project.
   B. Quality audits identify inefficiencies that exist within the processes and activities being performed.
   C. Quality audits ensure that the team members are doing their job and are necessary for documenting performance reviews.
   D. Conducting quality audits is a legal requirement.

9. Who is responsible for defining Theory X & Y?
   A. Douglas McGregor
   B. Frederick Herzberg
   C. Abraham Maslow
   D. Victor Vroom

10. The following are tools and techniques of the Manage Stakeholder Expectations process EXCEPT:
    A. Stakeholder register
    B. Communication methods
    C. Interpersonal skills
    D. Management skills

# Answers to Review Questions

1. B. Jon's upcoming role within the project has been pre-assigned. Be sure to keep your eyes open for clues in the question's scenario. For instance, Jon has been looking forward to this assignment for months, yet the project has not officially begun. This means that Jon would have been included within the project charter as a specific team member assigned to the role of software architect. Although C, project team member, is also correct, the question asks for the BEST answer.

2. C. Options A, B, and D are all benefits of virtual teams. A war room requires team members to be co-located, which means that they are based out of the same physical location. Since virtual teams are spread out in various locations, C is the correct choice.

3. B. Storming, the second of Bruce Tuckman and Mary Ann Jensen's five stages of team development, is when employees tend to be more confrontational with one another as they are vying for position and control.

4. D. The chaotic meeting scenario is a result of a lack of ground rules. Ground rules determine acceptable team behavior and rules that should be applied during meetings. As the scenario exhibits, a lack of ground rules can lead to poor productivity and waste of time.

5. A. The correct choice is social needs, which is the third level from the bottom in Maslow's pyramid. This is the level that describes the need to belong, be loved, and be accepted.

6. D. Confronting, also called problem solving, is the best strategy for resolving conflict because it brings about a win-win result.

7. C. Communication methods are a tool and technique of the Distribute Information process and include all means that make it possible to communicate project information to the appropriate individuals.

8. B. Quality audits bring about quality improvements and greater efficiency to how processes and activities are carried out, making B the correct choice. Option A describes process analysis. Although quality audits can reveal information on team performance, the purpose is not to conduct performance reviews. Quality audits ensure that applicable laws and standards are being adhered to, but it is not a legal requirement to conduct them.

9. A. Douglas McGregor is responsible for the Theory X & Y. Frederick Herzberg defined the Hygiene theory; Abraham Maslow created Maslow's hierarchy of needs theory; and Victor Vroom is responsible for the Expectancy theory.

10. A. The stakeholder register is an input to the process as opposed to a tool and technique. Avoid selecting an answer too fast simply because you recognize a connection between a term and a process.

Chapter

5

# Monitoring and Controlling the Project

## THE PMP EXAM CONTENT FROM THE MONITORING AND CONTROLLING THE PROJECT PERFORMANCE DOMAIN COVERED IN THIS CHAPTER INCLUDES THE FOLLOWING:

- ✓ Measure project performance using appropriate tools and techniques, in order to identify and quantify any variances, perform approved corrective actions, and communicate with relevant stakeholders.
- ✓ Manage changes to the project scope, schedule, and costs by updating the project plan and communicating approved changes to the team, in order to ensure that revised project goals are met.
- ✓ Ensure that project deliverables conform to the quality standards established in the quality management plan by using appropriate tools and techniques (e.g., testing, inspection, control charts), in order to satisfy customer requirements.
- ✓ Update the risk register and risk response plan by identifying any new risks, assessing old risks, and determining and implementing appropriate response strategies, in order to manage the impact of risks on the project.
- ✓ Assess corrective actions on the issue register and determine next steps for unresolved issues by using appropriate tools and techniques, in order to minimize the impact on project schedule, cost, and resources.
- ✓ Communicate project status to stakeholders for their feedback, in order to ensure the project aligns with business needs.

Monitoring and Controlling is the fourth of the five project management process groups and accounts for 25 percent of the questions on the PMP exam. The processes in the Monitoring and Controlling process group concentrate on monitoring and measuring project performance to identify variances from the project plan and get them back on track. Regularly monitoring project performance provides insight into the current state of the project and allows you to correct areas that are falling off track before they impact your project significantly. The Monitoring and Controlling process group is concerned not only with monitoring the project work but also with the project as a whole. In multiphase projects, this process group coordinates the project phases.

# Measuring Project Performance

As the project work is carried out, it is important to monitor the project performance closely. This can be done by using various tools and techniques that assist in identifying and quantifying variances as well as determining the necessary corrective action needed. Monitoring project performance occurs through the Monitor and Control Project Work process, part of the Project Integration Management Knowledge Area.

Work performed by sellers (resources external to the organization) are monitored and controlled through the Administer Procurements process. Administer Procurements is a process that belongs to the Project Procurement Management Knowledge Area.

## Monitor and Control Project Work

The Monitor and Control Project Work process involves monitoring all the processes in the Initiating, Planning, Executing, and Closing process groups to meet the performance objectives outlined in the project management plan. Collecting data, measuring results, and reporting on performance information are some of the activities performed during this process.

According to the *PMBOK® Guide*, the Monitor and Control Project Work process encompasses the following:

- Reporting and comparing actual project results against the project management plan
- Analyzing performance data and determining whether corrective or preventive action is needed

- Monitoring project risks
- Documenting all appropriate product information throughout the life of the project
- Gathering, recording, and documenting project information
- Monitoring approved change requests

Figure 5.1 shows the inputs, tools and techniques, and outputs of the Monitor and Control Project Work process.

**FIGURE 5.1** Monitor and Control Project Work process

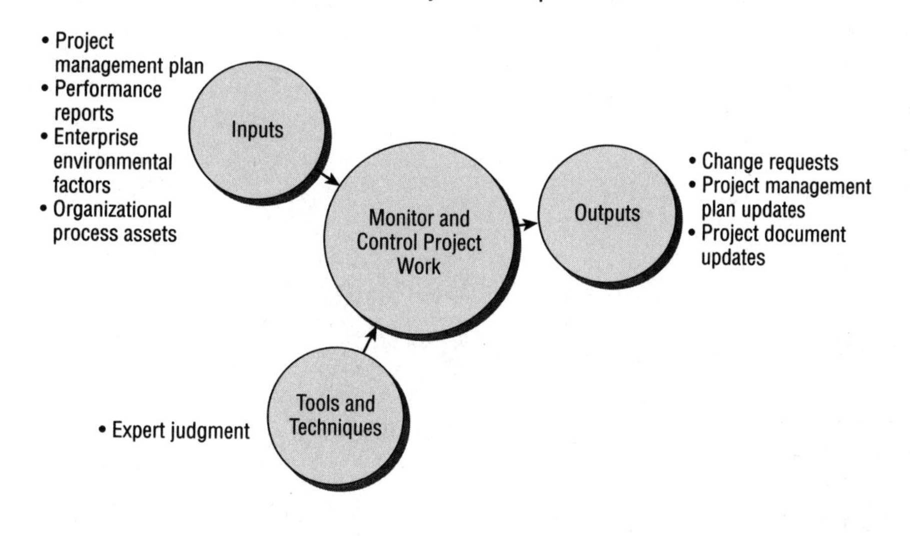

For more detailed information on the Monitor and Control Project Work process, see Chapter 10, "Measuring and Controlling Project Performance," in *PMP: Project Management Professional Exam Study Guide, 6th Edition* (Sybex, 2011).

## Inputs of Monitor and Control Project Work

Know the following inputs of the Monitor and Control Project Work process:

- Project management plan
- Performance reports
- Enterprise environmental factors
- Organizational process assets

**Project Management Plan** The project management plan provides the information necessary to monitor the project work and determine whether the project is progressing as planned.

**Performance Reports** Performance reports provide the current status and issues of the project.

**Enterprise Environmental Factors** The enterprise environmental factors provide the following information utilized by this process:

- Standards
- Work authorization systems
- Stakeholder risk tolerances
- Project management information systems

**Organizational Process Assets** The following organizational process assets are used as inputs:

- Communication requirements
- Procedures for financial controls, risk control, and problem solving
- Lessons learned database

## Tools and Techniques of Monitor and Control Project Work

Expert judgment is the one tool of the Monitor and Control Project Work process.

Expert judgment involves interpreting the information that results from this process.

## Outputs of Monitor and Control Project Work

The Monitor and Control Project Work process has three outputs.

**Change Requests** Change requests typically include the following items:

- Corrective actions
- Preventive actions
- Defect repairs

**Project Management Plan Updates** Project management plan updates may include updates to one or more of the following:

- Schedule management plan
- Cost management plan
- Quality management plan
- Scope baseline
- Schedule baseline
- Cost performance baseline

**Project Document Updates** Project document updates may include updates to the following:

- Forecasts
- Performance reports
- Issue log

## Administer Procurements

The Administer Procurements process is concerned with monitoring the vendor's performance and ensuring that all requirements of the contract are met. This process also manages the procurement relationships overall and makes changes and corrections as needed.

According to the *PMBOK® Guide*, you must integrate and coordinate the Direct and Manage Project Execution, Report Performance, Perform Quality Control, Perform Integrated Change Control, and Monitor and Control Risk processes during the Administer Procurements process.

Figure 5.2 shows the inputs, tools and techniques, and outputs of the Administer Procurements process.

**FIGURE 5.2** Administer Procurements process

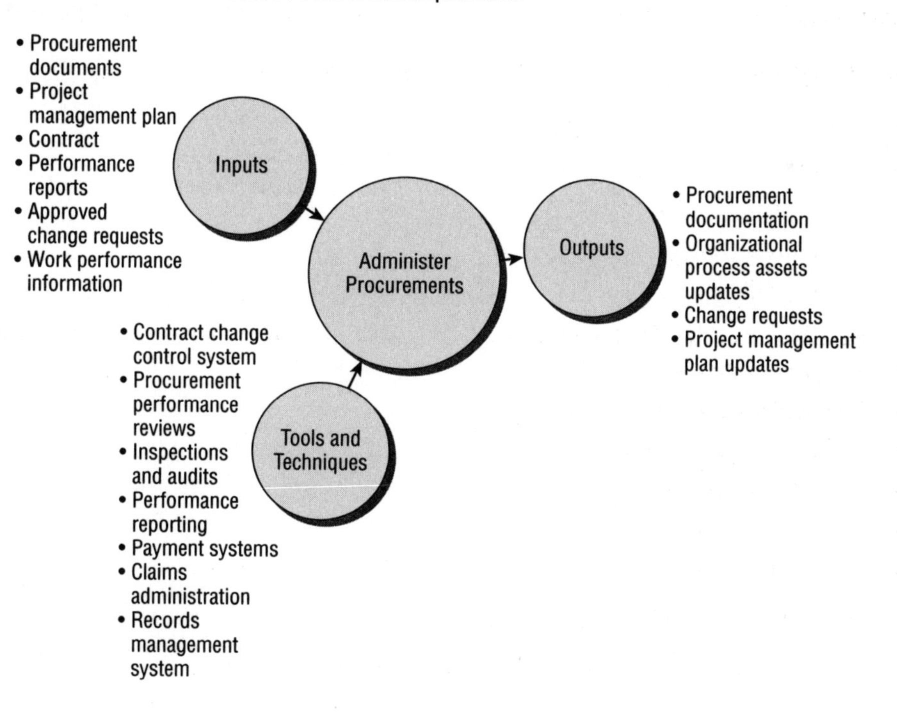

For more detailed information on the Administer Procurements process, see Chapter 10 in *PMP: Project Management Professional Exam Study Guide, 6th Edition.*

## Inputs of Administer Procurements

Know the following inputs of the Administer Procurements process:

- Procurement documents
- Project management plan
- Contract
- Performance reports
- Approved change requests
- Work performance information

**Procurement Documents** Procurement documents may include the following supporting documents:

- Procurement contract awards
- Statement of work

**Project Management Plan** The project management plan includes the procurement management plan, which guides the project management team in properly carrying out the process.

**Contract** To administer the relevant contracts, you must first include each contract as an input to this process.

**Performance Reports** Performance reports refer to seller-related documents, such as seller performance reports and technical documentation.

**Approved Change Requests** Approved change requests include changes that have been approved to the contract, such as these:

- Modifications to deliverables
- Changes to the product or service of the project
- Changes in contract terms
- Termination for poor performance

**Work Performance Information** Work performance information involves monitoring work results and examining the vendor's deliverables. This includes monitoring their work results against the project management plan and making sure that certain activities are performed correctly and in sequence.

## Tools and Techniques of Administer Procurements

Be familiar with the following tools and techniques of the Administer Procurements process:

- Contract change control system
- Procurement performance reviews
- Inspections and audits
- Performance reporting
- Payment systems
- Claims administration
- Records management system

**Contract Change Control System** The contract change control system describes the processes needed to make contract changes. Since the contract is a legal document, it cannot be changed without the agreement of all parties. It documents and includes the following items:

- Instructions for submitting changes
- Instructions for establishing the approval process
- Instructions outlining authority levels
- A tracking system for numbering the change requests and recording their status
- Procedures for dispute resolution

Figure 5.3 demonstrates where the contract change control system fits into the contract change process.

**FIGURE 5.3** Contract change control system

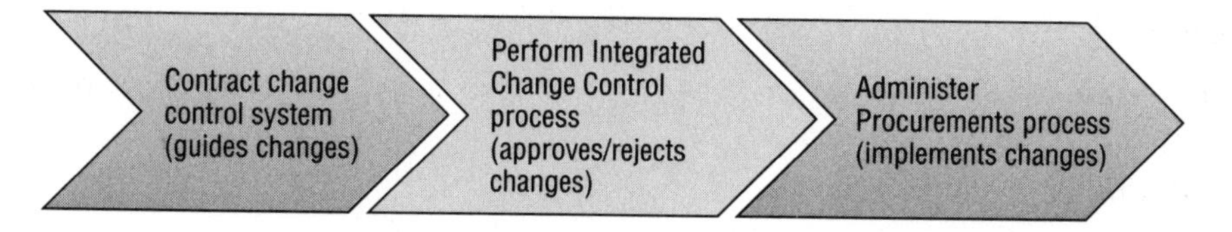


A formal process must be established to process and authorize (or deny) changes.

Authorization levels of the contract change control system are defined in the organizational policies.

**Procurement Performance Reviews** Procurement reviews examine the contract terms and seller performance. If the seller is not in compliance, action must be taken to either get them back into compliance or terminate the contract. These reviews can be conducted at the end of the contract's life cycle or at intervals during the contract period.

**Inspections and Audits** The purpose of inspections and audits is to determine whether there are any deficiencies in the seller's product or service. These are conducted by the buyer or a designated third party.

**Performance Reporting** Performance reporting involves providing managers and stakeholders with information about the vendor's progress in meeting the contract objectives.

**Payment Systems** Vendors submit seller invoices as an input to this process, and the payment system is the tool and technique used to issue payments.

**Claims Administration** Claims administration involves documenting, monitoring, and managing contested changes to the contract. Contested changes usually involve a disagreement about the compensation to the vendor for implementing the change. Here are some things to know about claims administration:

- Contested changes are also known as *disputes*, *claims*, or *appeals*.
- Claims can be settled directly between the parties themselves, through the court system, or by a process called arbitration.

According to the *PMBOK® Guide*, when the parties cannot reach an agreement themselves, they should use an alternative dispute-resolution process (ADR), like arbitration. Also note that the preferred method of settling disputes is negotiation.

**Records Management System** A records management system can be part of the project management information system (PMIS) and involves the following:

- Documentation
- Policies
- Control functions
- Automated tools

Records management systems typically index documents for easy filing and retrieval.

## Outputs of Administer Procurements

Know the following four outputs of the Administer Procurements process:

- Procurement documentation
- Organizational process assets updates

- Change requests
- Project management plan updates

**Procurement Documentation** Procurement documentation produced as part of this process includes, but is not limited to, the following:

- Contract
- Performance information
- Warranties
- Financial information
- Inspection and audit results
- Supporting schedules
- Approved and unapproved changes

**Organizational Process Assets Updates** Organizational process assets updates include the following items:

**Correspondence** Correspondence is information that needs to be communicated in writing to either the seller or the buyer, such as contract changes, audit results, notice of unsatisfactory performance, and notification of contract termination.

**Payment Schedules and Requests** Payment schedules and requests include information about paying the vendor on time and verifying that payment terms are met to warrant payment.

**Seller Performance Evaluation Documentation** Seller performance evaluation is a written record of the seller's performance on the contract and is prepared by the buyer. It should include information about whether the seller successfully met contract dates and fulfilled the requirements of the contract and/or contract statement of work and whether the work was satisfactory.

**Change Requests** Change requests may result from performing the Administer Procurements process and may include changes to the following documents:

- Project management plan
- Procurement management plan
- Cost baseline
- Project schedule

Change requests are submitted and processed through the integrated change control system.

**Project Management Plan Updates** Project management plan updates that may result from the Administer Procurements process typically include updates to the procurement management plan and baseline schedule.

**Exam Essentials**

**Name the outputs of the Monitor and Control Project Work process.** The outputs are change requests, which include corrective actions, preventive actions, and defect repair; project management plan updates; and project document updates.

**Name the processes that integrate with the Administer Procurements process.** Direct and Manage Project Execution, Report Performance, Perform Quality Control, Perform Integrated Change Control, and Monitor and Control Risk.

# Managing Changes to the Project Scope, Schedule, and Costs

Managing the triple constraints is an important part of monitoring and controlling a project. This includes managing changes to the project scope, schedule, and costs, which requires that the project plan be updated as needed. The Verify Scope, Control Scope, Control Schedule, and Control Costs processes focus on managing these changes and uncovering existing variance to the plan. As the plan is updated with approved changes, these changes should be communicated to the team to ensure that revised project goals are met.

## Verify Scope

Managing and reporting on project progress is the primary focus of the Monitoring and Controlling processes. One of the Monitoring and Controlling processes is Verify Scope. The primary purpose of the Verify Scope process is to formally accept completed deliverables and obtain sign-off that the deliverables are satisfactory and meet stakeholders' expectations. Verify Scope formalizes the acceptance of the project scope and is primarily concerned with the acceptance of work results.

Figure 5.4 shows the inputs, tools and techniques, and outputs of the Verify Scope process.

For more detailed information on the Verify Scope process, see Chapter 11, "Controlling Work Results," in *PMP: Project Management Professional Exam Study Guide, 6th Edition.*

**FIGURE 5.4** Verify Scope process

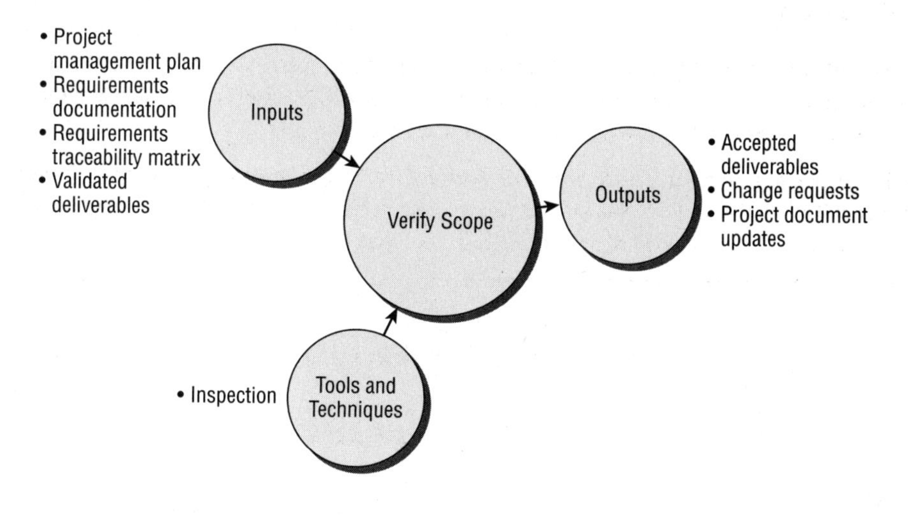


## Inputs of Verify Scope

Know the following inputs of the Verify Scope process:

- Project management plan
- Requirements documentation
- Requirements traceability matrix
- Validated deliverables

**Project Management Plan** The scope baseline is used in the project management plan. The scope baseline includes the project scope statement, work breakdown structure (WBS), and WBS dictionary, all of which will be utilized by this process.

**Requirements Documentation** To verify the scope of the project work, the project, product, and technical requirements (and any others) must be included as an input to this process.

**Requirements Traceability Matrix** The requirements traceability matrix tracks the requirements throughout the project life cycle.

**Validated Deliverables** Validated deliverables include deliverables that have been validated in the Perform Quality Control process.

You should perform Verify Scope even if the project is canceled, to document the degree to which the project was completed.

## Tools and Techniques of Verify Scope

Be familiar with the inspection tool and technique of the Verify Scope process. According to the *PMBOK® Guide*, inspection is concerned with making sure the project work and deliverables meet the requirements and product acceptance criteria. This may include ensuring, examining, and verifying activities.

## Outputs of Verify Scope

The Verify Scope process results in three outputs:

**Accepted Deliverables** The accepted deliverables output signifies that the deliverables have been formally accepted and signed off by the customer or sponsor.

**Change Requests** Change requests address deliverables that were rejected by the customer or sponsor and that therefore require defect repair.

**Project Document Updates** Project document updates typically include updates to the following items:

- Documents that define the product
- Documents that report completion status

# Control Scope

The Control Scope process involves monitoring the status of both the project and the product scope, monitoring changes to the project and product scope, and monitoring work results to ensure that they match expected outcomes. Any modification to the agreed-upon WBS is considered a scope change. This means the addition or deletion of activities or modifications to the existing activities on the WBS constitutes a project scope change.

Figure 5.5 shows the inputs, tools and techniques, and outputs of the Control Scope process.

**FIGURE 5.5** Control Scope process

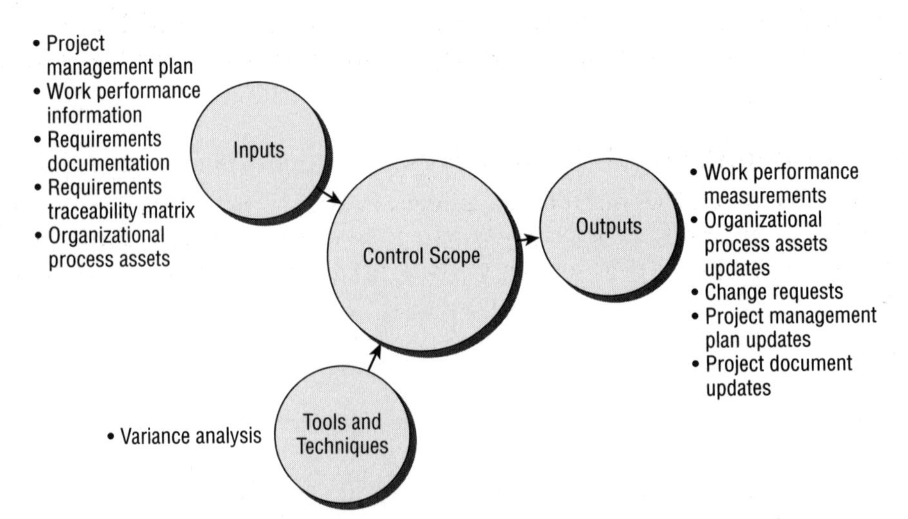

For more detailed information on the Control Scope process, see Chapter 11 in *PMP: Project Management Professional Exam Study Guide, 6th Edition.*

## Inputs of Control Scope

Know the following inputs of the Control Scope process:

- Project management plan
- Work performance information
- Requirements documentation
- Requirements traceability matrix
- Organizational process assets

**Project Management Plan** The following components of the project management plan are utilized in controlling scope:

  - Scope management plan
  - Scope baseline
  - Change management plan
  - Configuration management plan
  - Requirements management plan

**Work Performance Information** Work performance information provides the status and progress of deliverables.

**Requirements Documentation** Information from the requirements documentation input used in this process includes the project, product, technical and any other requirements.

**Requirements Traceability Matrix** The requirements traceability matrix is used to track the requirements throughout the project life cycle.

**Organizational Process Assets** Organizational process assets utilized in this process include any policies, procedures, guidelines, and reporting methods that relate to scope control.

## Tools and Techniques of Control Scope

The Control Scope process has one tool and technique: variance analysis. Variance analysis includes reviewing project performance measurements to determine whether there are variances in project scope and whether any corrective action is needed as a result of existing variances.

Unapproved or undocumented changes that sometimes make their way into the project are referred to as scope creep.

### Outputs of Control Scope

The Control Scope process results in the following outputs:

- Work performance measurements
- Organizational process assets updates
- Change requests
- Project management plan updates
- Project document updates

**Work Performance Measurements** Work performance measurements compare planned and actual scope performance.

**Organizational Process Assets Updates** The following updates are typically made to the organizational process assets:

- Causes of variances
- Corrective actions taken
- Lessons learned

**Change Requests** When scope changes are requested, all areas of the project should be investigated to determine what the changes would impact. Change requests typically involve the scope baseline or components of the project management plan.

**Project Management Plan Updates** Depending on the approved changes, updates to the project management plan may result in changes to the scope baseline or other project baselines.

**Project Document Updates** Updates to project documents that occur as a result of this process typically include the requirements documentation and requirements traceability matrix.

## Control Schedule

The Control Schedule process involves determining the status of the project schedule, determining whether changes have occurred or should occur, and influencing and managing schedule changes.

Figure 5.6 shows the inputs, tools and techniques, and outputs of the Control Schedule process.

**FIGURE 5.6** Control Schedule process

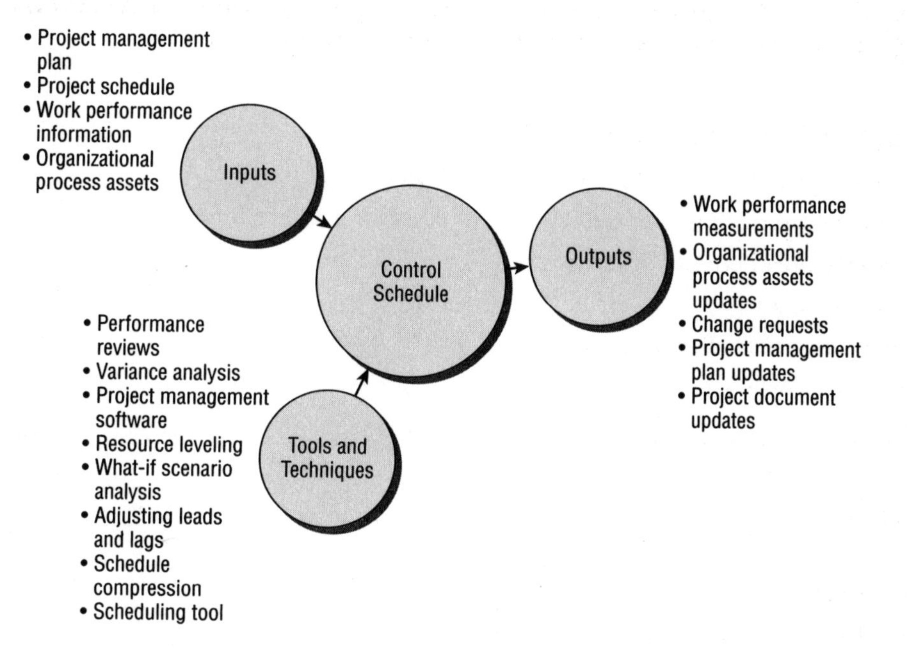

For more detailed information on the Control Schedule process, see Chapter 11 in *PMP: Project Management Professional Exam Study Guide, 6th Edition.*

## Inputs of Control Schedule

Know the following inputs of the Control Schedule process:

- Project management plan
- Project schedule
- Work performance information
- Organizational process assets

**Project Management Plan** The schedule management plan and the schedule baseline are utilized to carry out the Control Schedule process from within the project management plan.

**Project Schedule** To control the schedule, the project schedule itself must be included as an input into the process.

**Work Performance Information** Work performance information provides status information on the activities.

**Organizational Process Assets** The following organizational process assets are utilized in this process:

- Policies, procedures, guidelines, and reporting methods relating to schedule control
- Schedule control tools

## Tools and Techniques of Control Schedule

The Control Schedule process includes the following tools and techniques:

- Performance reviews
- Variance analysis
- Project management software
- Resource leveling
- What-if scenario analysis
- Adjusting leads and lags
- Schedule compression
- Scheduling tool

**Performance Reviews** Performance reviews in this process examine elements such as actual start and end dates for schedule activities and the remaining time to finish uncompleted activities. Schedule variance (SV) and schedule performance index (SPI) can be used to determine the impact of the schedule variations and whether corrective action is necessary.

**Variance Analysis** SV and SPI calculations can be used to determine how the actual schedule varies from the schedule baseline.

**Project Management Software** Project management scheduling software can be used to track actual versus planned activity dates and the impact of any existing changes.

**Resource Leveling** Resource leveling is used to better distribute the work among resources, literally leveling out the use of the resources.

**What-If Scenario Analysis** What-if scenarios are used to bring the actual schedule back in line with the planned schedule.

**Adjusting Leads and Lags** Leads and lags are used to bring the actual schedule back in line with the planned schedule.

**Schedule Compression** Schedule compression techniques, such as fast-tracking and crashing, are used to bring the actual schedule back in line with the planned schedule.

**Scheduling Tool** Scheduling tools are used to perform schedule network analysis, which results in an updated project schedule.

## Outputs of Control Schedule

The Control Schedule process includes the following outputs:

- Work performance measurements
- Organizational process assets updates
- Change requests
- Project management plan updates
- Project document updates

**Work Performance Measurements** Work performance measurements include the calculated SV and SPI values.

**Organizational Process Assets Updates** Organizational process assets updates typically include the following items:

- Causes of variances
- Corrective actions
- Lessons learned

**Change Requests** Changes requests generated as a result of this process typically include modifications to the schedule baseline.

**Project Management Plan Updates** Project management plan updates typically include updates to the schedule baseline, the cost baseline, and the schedule management plan.

**Project Document Updates** The project document updates output may require updates to the schedule data or the project schedule.

# Control Costs

The Control Costs process monitors the project budget, manages changes to the cost performance baseline, and records actual costs. It's concerned with monitoring project costs to prevent unauthorized or incorrect costs from being included in the cost baseline. The activities are among those included in this process:

- Monitoring changes to costs or the cost performance baseline and understanding variances from the baseline
- Monitoring change requests that affect cost and resolving them in a timely manner
- Informing stakeholders of approved changes and their costs

Figure 5.7 shows the inputs, tools and techniques, and outputs of the Control Costs process.

**FIGURE 5.7** Control Costs process

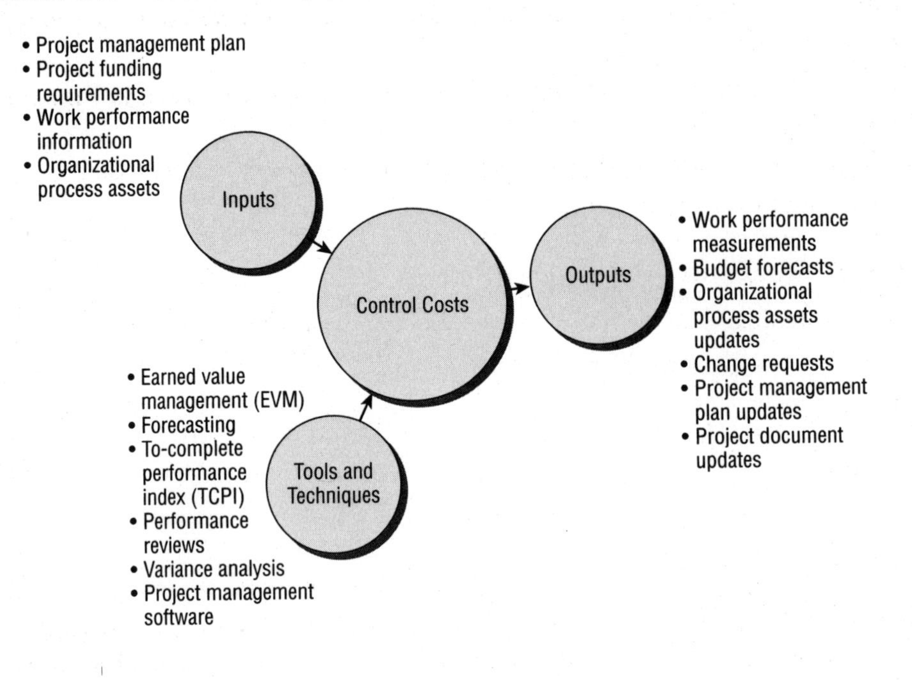

For more detailed information on the Control Costs process, see Chapter 11 in *PMP: Project Management Professional Exam Study Guide, 6th Edition.*

## Inputs of Control Costs

Know the following inputs of the Control Costs process:

- Project management plan
- Project funding requirements
- Work performance information
- Organizational process assets

**Project Management Plan** The project management plan includes the cost performance baseline and the cost management plan, which will be utilized by this process. The cost performance baseline compares actual expenditures to date on the project to the baseline. The cost management plan details how costs should be monitored and controlled throughout the life of the project.

**Project Funding Requirements** Project funding requirements include the periodic and total project funding provided.

**Work Performance Information** The work performance information input provides information about the project's progress, including authorized and incurred costs and project estimates.

**Organizational Process Assets** The following organizational process assets are utilized by this process:

- Policies, procedures, guidelines, and reporting methods relating to cost
- Cost control tools

## Tools and Techniques of Control Costs

You should be familiar with the following tools and techniques of the Control Costs process:

- Earned value management (EVM)
- Forecasting
- To-complete performance index (TCPI)
- Performance reviews
- Variance analysis
- Project management software

**Earned Value Management** Performance measurement analysis can be accomplished by using a technique called earned value management (EVM). EVM compares what you've received or produced to what you've spent and establishes the cause and impact of variances to determine necessary corrective action. It is the most often used performance measurement method.

EVM is performed on the work packages and the control accounts of the WBS. To perform the EVM calculations, the planned value (PV), earned value (EV), and actual cost (AC) are collected. They are defined in Table 5.1 along with other terms associated with the EVM technique.

**TABLE 5.1** Earned value management definitions

| Term | Definition |
|---|---|
| Planned value (PV) | The PV is the budgeted cost of work that has been authorized for a schedule activity or WBS component during a given time period or phase. These budgets are established during the Planning processes. All PVs add up to the budget at completion (BAC). PV is also known as budgeted cost of work scheduled (BCWS). |
| Earned value (EV) | EV is measured as budgeted dollars for the work performed. EV is typically expressed as a percentage of the work completed compared to the budget. It is also known as budgeted cost of work performed (BCWP). |

**TABLE 5.1** Earned value management definitions *(continued)*

| Term | Definition |
|---|---|
| Actual cost (AC) | AC is the cost of completing the work component in a given time period. AC measures the costs (direct, indirect, or other) that were used to calculate the planned value. It is also known as actual cost of work performed (ACWP). |
| Cost variance (CV) | CV determines whether costs are higher or lower than budgeted. A negative CV means the project is over budget; a positive CV means the project is under budget. |
| Cost performance index (CPI) | CPI measures the value of the work completed against actual cost. A CPI of 1 means the project is on budget; a CPI >1 means the project is under budget; a CPI <1 means the project is over budget. |
| Schedule variance (SV) | SV determines whether the work is ahead of or behind the planned schedule. A positive SV means the project is ahead of schedule; a negative SV means the project is behind schedule. |
| Schedule performance index (SPI) | SPI measures the schedule progress to date against the planned progress. An SPI of 1 means the project is on schedule, an SPI of >1 means the project is ahead of schedule, and an SPI of <1 means the project is behind schedule. |
| Estimate at completion (EAC) | EAC is the expected total cost of a work component, a schedule activity, or the project at its completion. |
| Estimate to complete (ETC) | ETC is the amount of effort remaining based on completed and expected activities. |
| Budget at completion (BAC) | BAC is the sum of all the budgets (or PVs) established. |
| Variance at completion (VAC) | VAC is the difference between the budget at completion and the estimate at completion. |

PV, AC, and EV measurements can be plotted graphically to show the variances between them. If there are no variances in the measurements, all lines on the graph remain the same, meaning that the project is progressing as planned. All of these measurements include a cost component. Costs are displayed in an S curve because spending is minimal in the beginning of the project, picks up steam toward the middle, and then tapers off at the end. This means your earned value measurements will also take on the S curve shape. Figure 5.8 shows an example that plots these three measurements.

**FIGURE 5.8** Earned value

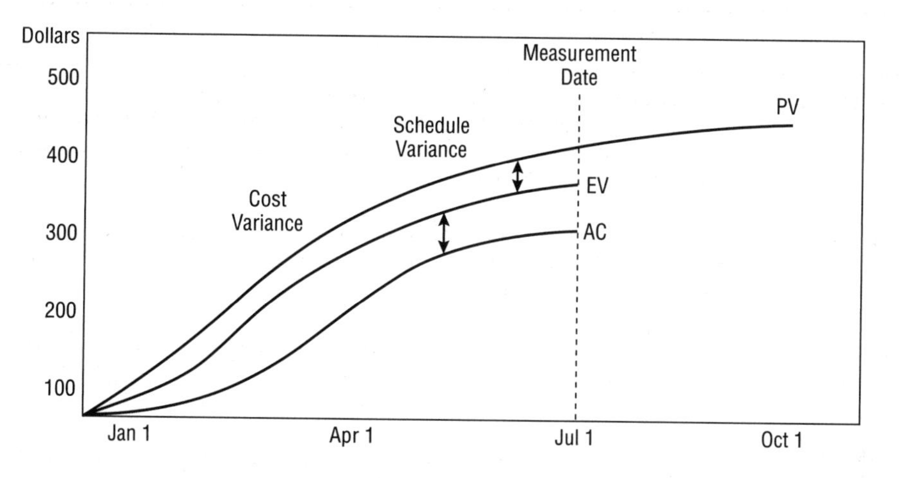


**Cost Variance** Cost variance is one of the most popular variances that project managers use, and it tells you whether your costs are higher than budgeted (with a resulting negative number) or lower than budgeted (with a resulting positive number). It measures the actual performance to date or during the period against what's been spent.

Use this formula:

$$CV = EV - AC$$

**Schedule Variance** Schedule variance, another popular variance, tells you whether the schedule is ahead or behind what was planned for this period. This formula is most helpful when you've used the critical path methodology to build the project schedule.

Use this formula:

$$SV = EV - PV$$

Together, the $CV$ and $SV$ are known as efficiency indicators for the project and can be used to compare performance of all the projects in a portfolio.

Cost and schedule performance indexes are primarily used to calculate performance efficiencies, and they're often used to help predict future project performance.

**Cost Performance Index** The cost performance index (CPI) measures the value of the work completed against actual cost. It is the most critical of all the EVM measurements according to the *PMBOK® Guide* because it tells you the cost efficiency for the work completed to date or at the completion of the project. As mentioned previously, if CPI is greater than 1, your spending is less than anticipated. If CPI is less than 1, you are spending more than anticipated for the work completed.

Use this formula:

$$CPI = EV \div AC$$

**Schedule Performance Index** The schedule performance index (SPI) measures the progress to date against the progress that was planned. This formula should be used in conjunction with an analysis of the critical path activities to determine if the project will finish ahead of or behind schedule. As mentioned previously, if SPI is greater than 1, you are ahead of schedule. If SPI is less than 1, you are behind schedule.

Use this formula:

$$SPI = EV \div PV$$

**Forecasting** Forecasting uses the information gathered to date and estimates the future conditions or performance of the project based on what is known when the calculation is performed.

One of the forecasting formulas used is called estimate at completion (EAC). EAC estimates the expected total cost of a work component, a schedule activity, or the project at its completion. This is the probable final value for the work component (or project).

Use this formula:

$$EAC = AC + \textit{bottom-up } ETC$$

The bottom-up estimate to complete (ETC) is the amount of effort remaining based on the activities completed to date and what is thought to occur in the future. Each estimate is summed to come up with a total ETC.

According to the *PMBOK® Guide*, there are three EAC forecasting formulas (see Table 5.2), which use budget at completion (BAC). BAC is the sum of all the budgets established for all the work in the work package, control account, schedule activity, or project. It's the total planned value for the work component or project. In addition to the bottom-up ETC, there are two other formulas for calculating ETC that you should be aware of for the exam. These are also listed in Table 5.2.

**TABLE 5.2** EAC forecasting

| Forecast | Formula | Description |
|---|---|---|
| For ETC work performed at the budgeted rate | $EAC = AC + BAC - EV$ | This formula calculates EAC based on the actual costs to date and the assumption that ETC work will be completed at the budgeted rate, thereby accepting previous performance (whether good or bad). |
| For ETC work performed at the present CPI | $EAC = BAC \div CPI$ | This forecast assumes that future performance will be just like the past performance for the project. |

| Forecast | Formula | Description |
|---|---|---|
| For ETC work considering both SPI and CPI factors | $EAC = AC + ((BAC - EV) \div (CPI \times SPI))$ | This formula assumes there is a negative cost performance to date and the project schedule dates must be met. |
| For future cost variances that will be similar | $ETC = (BAC - EV) \div CPI$ | Use this formula for predicting future cost variances that will be similar to the types of variances you've seen to date. |
| For future cost variances that will *not* be similar | $ETC = BAC - EV$ | Use this formula for predicting future cost variances that will *not* be similar to the types of variances you've seen to date. |

**To-Complete Performance Index** To-complete performance index (TCPI) is the projected performance level the remaining work of the project must achieve to meet the BAC or EAC. It's calculated by dividing the work that's remaining by the funds that are remaining.

The formula for TCPI when using the BAC is as follows:

$$TCPI = (BAC - EV) \div (BAC - AC)$$

If the result of calculating the TCPI is less than 1, future work does not have to be performed as efficiently as past performance.

When the BAC is no longer attainable, the project manager should calculate a new EAC, and this new estimate becomes the goal you'll work toward once it's approved by management. The formula for TCPI when EAC is your goal is as follows:

$$TCPI = (BAC - EV) \div (EAC - AC)$$

If cumulative CPI falls below 1, all future project work must be performed at the TCPI.

**Performance Reviews** Performance reviews compare cost performance over time and the estimates of funds needed to complete the remaining work. The following three analysis types are associated with performance reviews:

- Variance analysis calculates the difference between the budget at completion and the estimate at completion. To calculate this, use the variance at completion (VAC) formula:

  $$VAC = BAC - EAC$$

- Trend analysis analyzes existing trends of project performance over time and whether performance is improving or declining.
- Earned value performance includes a comparison of the baseline plans to actual performance.

**Variance Analysis** Variance analysis works closely with the performance reviews technique. Aside from calculating VAC, variance analysis also focuses on the cause that led to a deviation from the baselines. By determining the cause of existing or potential variance, the project management team can then formulate an effective response through a change request.

**Project Management Software** Project management software is used to monitor PV, EV, and AC. Table 5.3 shows the formulas used to calculate values.

**TABLE 5.3** Earned value formulas

| Item | Formula | Comments |
|---|---|---|
| Cost variance (CV) | $CV = EV - AC$ | |
| Schedule variance (SV) | $SV = EV - PV$ | |
| Cost performance index (CPI) | $CPI = EV \div AC$ | |
| Schedule performance index (SPI) | $SPI = EV \div PV$ | |
| Estimate at completion (EAC) | $EAC = AC + \text{bottom-up } ETC$ | EAC using actual costs to date plus new budgeted rates based on expert feedback. |
| Estimate at completion (EAC) | $EAC = AC + BAC - EV$ | EAC using actual costs to date and future performance will be based on budgeted rate. |
| Estimate at completion (EAC) | $EAC = BAC \div CPI$ | EAC assuming future performance will behave like past performance. |
| Estimate at completion (EAC) | $EAC = AC + [(BAC - EV) \div (CPI \times SPI)]$ | EAC when cost performance is negative and schedule dates must be met. |
| Estimate to complete (ETC) | $ETC = (BAC - EV) \div CPI$ | When future cost variances will be similar to the types of variances experienced to date. |
| Estimate to complete (ETC) | $ETC = (BAC - EV)$ | When future cost variances will *not* be similar to the variances experienced to date. |

| Item | Formula | Comments |
|---|---|---|
| To-complete performance index (TCPI) | $TCPI = (BAC - EV) \div (BAC - AC)$ | This TCPI formula is used when employing BAC. |
| To-complete performance index (TCPI) | $TCPI = (BAC - EV) \div (EAC - AC)$ | This TCPI formula is used when EAC is your goal. |
| Variance at completion (VAC) | $VAC = BAC - EAC$ | |

## Outputs of Control Costs

The Control Costs process results in the following outputs:

- Work performance measurements
- Budget forecasts
- Organizational process assets updates
- Change requests
- Project management plan updates
- Project document updates

**Work Performance Measurements** Work performance measurements include the calculated CV, SV, CPI, and SPI values.

**Budget Forecasts** Budget forecasts include the calculated EAC, ETC, TCPI, and VAC.

**Organizational Process Assets Updates** Organizational process assets updates typically include the following items:

- Causes of variances
- Corrective actions taken
- Lessons learned

**Change Requests** Change requests that result from this process typically involve a change to the cost performance baseline.

**Project Management Plan Updates** Project management plan updates typically include updates to the cost performance baseline and cost management plans, which should reflect any approved changes.

**Project Document Updates** Project document updates typically include cost estimates and the basis of estimates.

**Exam Essentials**

**Name the purpose of the Verify Scope process.** The purpose of Verify Scope is to determine whether the work is complete and whether it satisfies the project objectives.

**Describe the purpose of the Control Costs process.** The Control Costs process is concerned with monitoring project costs to prevent unauthorized or incorrect costs from being included in the cost baseline.

**Be able to describe earned value measurement techniques.** Earned value measurement (EVM) monitors the planned value (PV), earned value (EV), and actual cost (AC) expended to produce the work of the project. Cost variance (CV), schedule variance (SV), cost performance index (CPI), and schedule performance index (SPI) are the formulas used with the EVM technique.

**Know the purpose of the Control Scope and Control Schedule processes.** The purpose of the Control Scope process is to monitor project and product scope and manage changes to the scope baseline. The purpose of the Control Schedule process is to monitor the status of the project's progress and to manage changes to the schedule baseline.

**Be able to name the tools and techniques of the Control Costs process.** The tools and techniques of Control Costs are earned value management, forecasting, to-complete performance index, performance reviews, variance analysis, and project management software.

# Ensuring Adherence to Quality Standards

The Perform Quality Control process is responsible for ensuring that project deliverables conform to the quality standards as laid out in the quality management plan. This inspection can be carried out by using several tools and techniques, such as control charts and testing, in order to satisfy customer requirements.

## Perform Quality Control

The Perform Quality Control process identifies the causes of poor product quality or processes and makes recommendations to bring them up to the required levels. Quality control is practiced throughout the project to identify and remove the causes of unacceptable results.

Figure 5.9 shows the inputs, tools and techniques, and outputs of the Perform Quality Control process.

**FIGURE 5.9** Perform Quality Control process

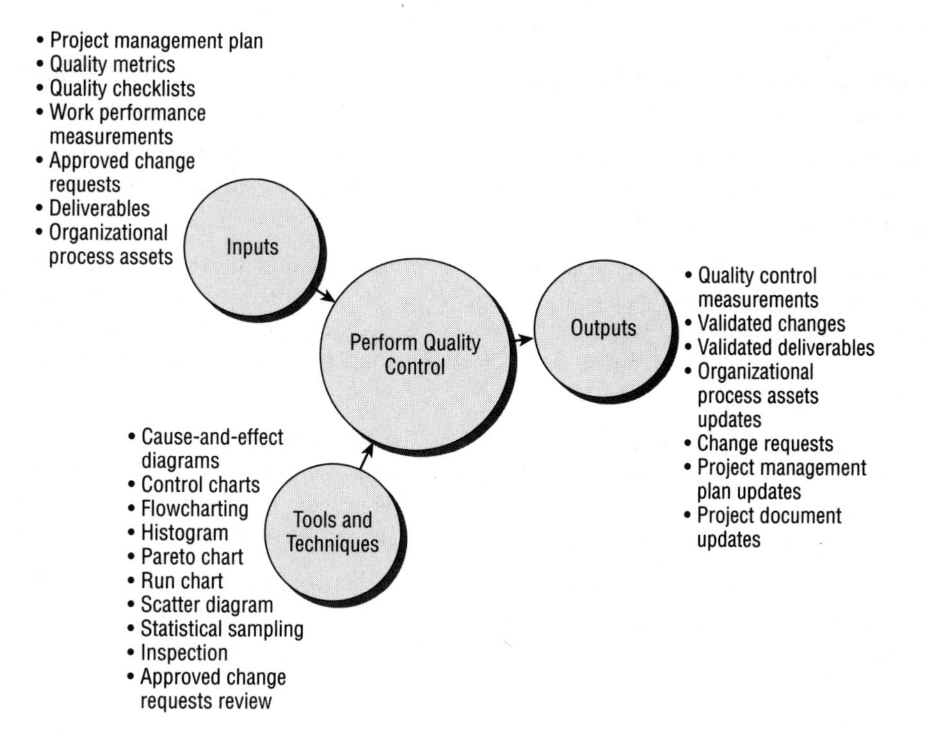

For more detailed information on the Perform Quality Control process, see Chapter 11 in *PMP: Project Management Professional Exam Study Guide, 6th Edition.*

## Inputs of Perform Quality Control

Know the following inputs of the Perform Quality Control process:

- Project management plan
- Quality metrics
- Quality checklists
- Work performance measurements
- Approved change requests
- Deliverables
- Organizational process assets

**Project Management Plan** The project management plan contains the quality management plan, which is used to carry out the Perform Quality Control process.

**Quality Metrics** Quality metrics provide the basis for measuring project or product quality attributes in this process.

**Quality Checklists** Quality checklists in this process are used to ensure that the set of required steps determined in the Plan Quality process are completed.

**Work Performance Measurements** Work performance measurements allow the project team to evaluate actual versus planned progress in the areas of technical performance, schedule performance, and cost performance.

**Approved Change Requests** Approved change requests are requested changes that were approved through the Perform Integrated Change Control process and submitted as an input for implementation.

**Deliverables** Part of achieving quality is ensuring that all deliverables are completed successfully. Deliverables will be validated as part of this process.

**Organizational Process Assets** The following organizational process assets are utilized by this process:

- Standards, policies, and guidelines relating to quality standards
- Procedures for reporting and communicating issues and defects

## Tools and Techniques of Perform Quality Control

Be familiar with the following tools and techniques of the Perform Quality Control process and how they are utilized in this process:

- Cause-and-effect diagrams
- Control charts
- Flowcharting
- Histogram
- Pareto chart
- Run chart
- Scatter diagram
- Statistical sampling
- Inspection
- Approved change requests review

The first seven tools and techniques of the Perform Quality Control process are collectively known as the seven basic tools of quality.

**Cause-and-Effect Diagrams** Cause-and-effect diagrams, also known as *Ishikawa* or *fishbone diagrams*, help identify root causes of issues. The idea is to determine the possible issues or effects of various factors. A cause and its potential sub-causes are plotted on the diagram, which resembles the skeleton of a fish. Figure 5.10 shows what the cause-and-effect diagram may look like.

**FIGURE 5.10** Cause-and-effect diagram

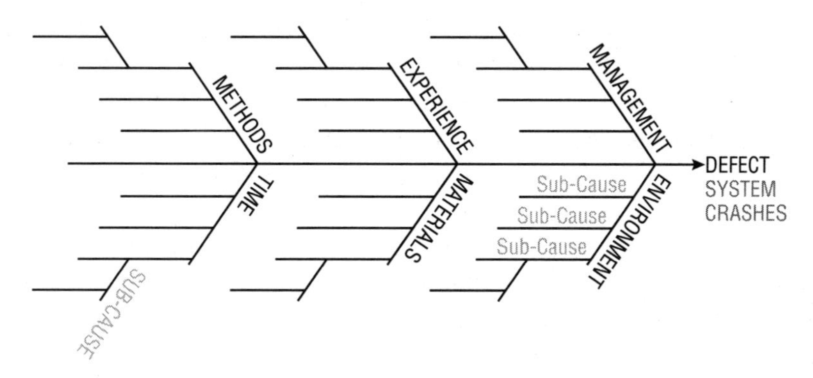


**Control Charts** Control charts measure the results of processes over time and display the results in graph form. They provide a way of measuring variances to determine whether process variances are in or out of control.

A control chart is based on sample variance measurements. From the samples chosen and measured, the mean and standard deviation are determined. Quality is usually maintained—or said to be in control—in plus or minus three standard deviations, also referred to as the upper and lower control limits.

Figure 5.11 illustrates a sample control chart.

**FIGURE 5.11** Control chart

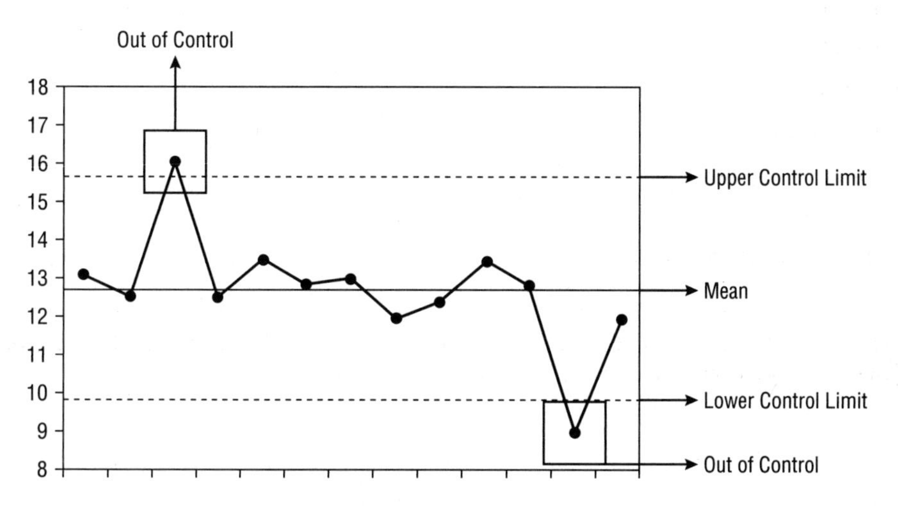

Out-of-control variances should be investigated to determine whether corrective action is needed.

For reference, know that standard deviation, or sigma, is equal to the following levels of quality:

±1 sigma is equal to 68.26 percent

±2 sigma is equal to 95.46 percent

±3 sigma is equal to 99.73 percent

±6 sigma is equal to 99.99 percent

Perform Quality Control says that if the process measurements fall in the control limits, then 99.73 percent (3 sigma) of the parts going through the process will fall in an acceptable range of the mean.

**Flowcharting** Flowcharts are diagrams that show the logical steps that must be performed to accomplish an objective; they can also show how the individual elements of a system interrelate. Flowcharting helps identify where quality problems might occur on the project and how problems occur.

**Histogram** Histograms are typically bar charts that depict the distribution of variables over time. In Perform Quality Control, the histogram usually depicts the attributes of the problem or situation.

**Pareto Chart** Pareto charts are named after Vilfredo Pareto, who is credited for discovering the theory behind them, which is based on the *80/20 rule*. The 80/20 rule as it applies to quality states that a small number of causes (20 percent) create the majority of the problems (80 percent). Pareto theorized that you get the most benefit by spending the majority of time fixing the most important problems.

Pareto charts are displayed as histograms that rank-order the most important factors—such as delays, costs, and defects—by their frequency over time. The information shown in Table 5.4 is plotted on an example Pareto chart shown in Figure 5.12.

**TABLE 5.4** Frequency of failures

| Item | Defect Frequency | Percent of Defects | Cumulative Percent |
|---|---|---|---|
| A | 800 | .33 | .33 |
| B | 700 | .29 | .62 |
| C | 400 | .17 | .79 |
| D | 300 | .13 | .92 |
| E | 200 | .08 | 1.0 |

**FIGURE 5.12** Pareto chart

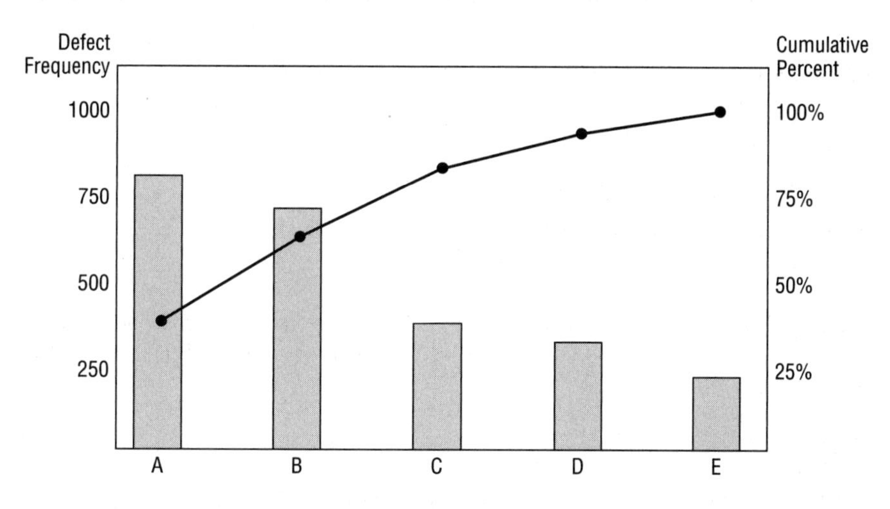


The problems are rank-ordered according to their frequency and percentage of defects. The defect frequencies in this figure appear as black bars, and the cumulative percentages of defects are plotted as circles. The rank ordering of these problems demonstrate where corrective action should be taken first.

**Run Chart** Run charts are used to show variations or trends in the process over time. They are similar to control charts, but without a displayed limit. Here's some additional information to know about run charts:

- Common causes of variances come about as a result of circumstances or situations that are relatively common to the process you're using and are easily controlled at the operational level. There are three types of variances that make up common causes of variances:
  - Random variances
  - Known or predictable variances
  - Variances that are always present in the process
- Special-cause variances are variances that are not common to the process.

Trend analysis is another technique that's carried out by using run charts. Trend analysis in the Perform Quality Control process is a mathematical technique that uses historical results to predict future outcomes, such as these:

- Technical performance, which compares the technical accomplishments of project milestones completed to the technical milestones defined in the project Planning process group
- Cost and schedule performance, which compares the number of variance activities occurring per period

**Scatter Diagram** Scatter diagrams use two variables, one called an independent variable, which is an input, and one called a dependent variable, which is an output. Scatter diagrams display the relationship between these two elements as points on a graph. This relationship is typically analyzed to prove or disprove cause-and-effect relationships.

Figure 5.13 shows a sample scatter diagram.

**FIGURE 5.13** Scatter diagram

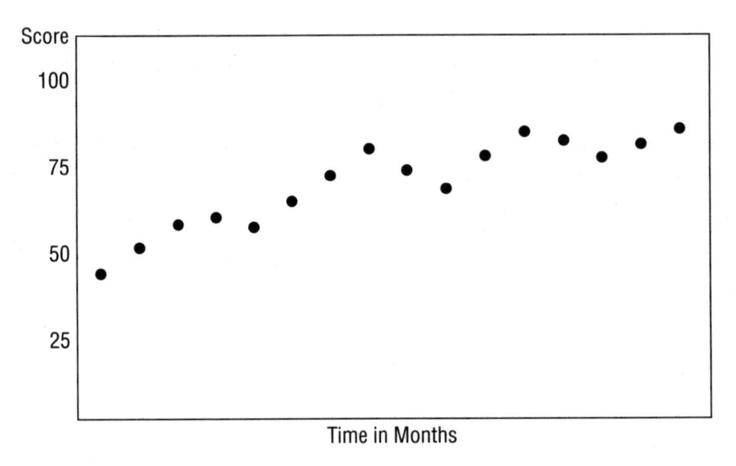


**Statistical Sampling** Statistical sampling involves taking a sample number of parts from the whole population and examining them to determine whether they fall in acceptable variances. The quality plan defines how to measure the results.

**Inspection** Inspection involves physically looking at, measuring, or testing results to determine whether they conform to the requirements or quality standards. Inspection will tell you where problems exist and gives you the opportunity to correct them, thus leading to quality improvements. Here's some additional information to know about inspections:

- Inspections might occur after the final product is produced or at intervals during the development of the product to examine individual components.
- Inspections are also known as *reviews* and *peer reviews.*
- Measurements that fall in a specified range are called tolerable results.
- One inspection technique uses measurements called attributes. Attribute sampling is used to verify conformance and nonconformance to predefined specifications.

**Approved Change Requests Review** Change requests that have been approved and implemented are reviewed to ensure that they were implemented correctly.

## Outputs of Perform Quality Control

The Perform Quality Control process results in the following outputs:

- Quality control measurements
- Validated changes

- Validated deliverables
- Organizational process assets updates
- Change requests
- Project management plan updates
- Project document updates

**Quality Control Measurements** Quality control measurements include the results of the quality control activities performed during this process.

**Validated Changes** Validated changes are the results of changes, defect repairs, or variances that have been inspected and corrected.

**Validated Deliverables** Validated deliverables are deliverables that have been taken through the tools and techniques of this process to determine if they are correct and accurate.

**Organizational Process Assets Updates** Organizational process assets updates include the following items:

- Completed checklists
- Lessons-learned documentation

**Change Requests** Change requests to the project management plan may result from carrying out this process and will be submitted to the Perform Integrated Change Control process.

**Project Management Plan Updates** Updates to the project management plan typically consist of changes to the quality management plan and the process improvement plan.

**Project Document Updates** Project document updates typically include updates to quality standards.

**Exam Essentials**

**Describe the purpose of the Perform Quality Control process.** The purpose of the Perform Quality Control process is to monitor work results to see whether they comply with the standards set out in the quality management plan.

# Updating the Risk Register

Identifying new risk, assessing old risks, and implementing the appropriate response strategies should occur as an integral part of monitoring and controlling activities. This is important to managing the impact of risks on the project and is carried out through the Monitor and Control Risks process.

## Monitor and Control Risks

The Monitor and Control Risks process should be carried out throughout the project life cycle. In addition to the activities just mentioned, this process is responsible for improving the risk management processes in a number of ways:

- Evaluating risk response plans that are put into action as a result of risk events
- Monitoring the project for risk triggers
- Reexamining existing risks to determine if they have changed or should be closed out
- Monitoring residual risks
- Reassessing project assumptions and determining validity
- Ensuring that policies and procedures are followed
- Ensuring that risk response plans and contingency plans are put into action appropriately and are effective
- Ensuring that contingency reserves (for schedule and cost) are updated according to the updated risk assessment
- Evaluating the overall effectiveness of the risk processes

Figure 5.14 shows the inputs, tools and techniques, and outputs of the Monitor and Control Risks process.

**FIGURE 5.14** Monitor and Control Risks process

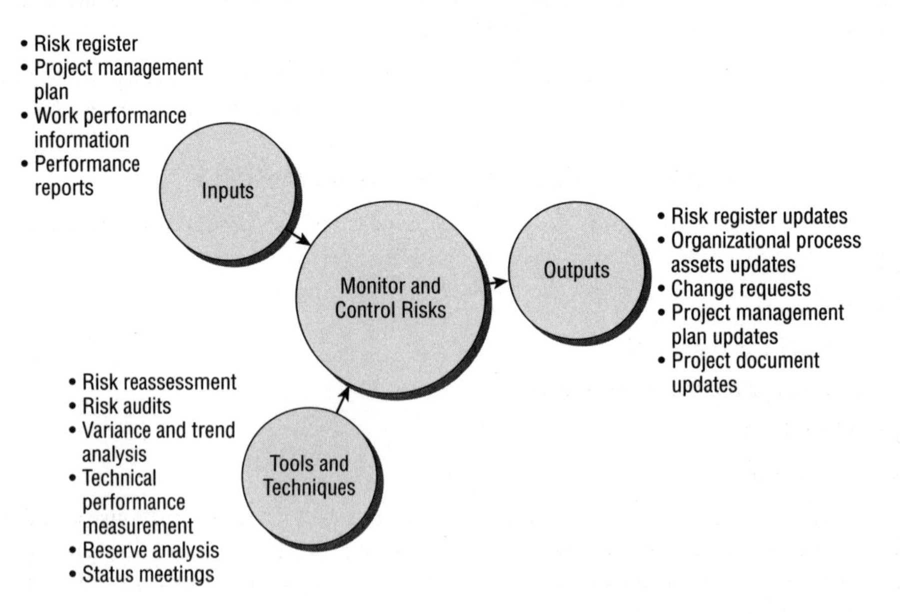

For more detailed information on the Monitor and Control Risks process, see Chapter 11 in *PMP: Project Management Professional Exam Study Guide, 6th Edition.*

## Inputs of Monitor and Control Risks

There are four inputs of the Monitor and Control Risks process that you should know:

- Risk register
- Project management plan
- Work performance information
- Performance reports

**Risk Register** The risk register contains all of the documented information of the identified risks. It tracks and ranks individual risks, identifies the risk owner, describes risk triggers and residual risks, and lists the response plans and strategies you should implement in the event of an actual risk event.

**Project Management Plan** The project management plan includes the risk management plan, which provides the necessary information for carrying out this process.

**Work Performance Information** Work performance information is used in monitoring previously identified risks. It also includes information that may assist in the identification of risk triggers, indicating that a risk event is about to occur.

**Performance Reports** Performance reports are examined from the perspective of risk and include information such as status reports, performance measurements, and forecasts.

## Tools and Techniques of Monitor and Control Risks

Know the following tools and techniques of the Monitor and Control Risks process:

- Risk reassessment
- Risk audits
- Variance and trend analysis
- Technical performance measurement
- Reserve analysis
- Status meetings

**Risk Reassessment** Periodic, scheduled reviews of identified risks, risk responses, and risk priorities should occur during the project. The idea is to monitor risks and their status and determine whether their consequences still have the same impact on the project objectives as when they were originally planned.

**Risk Audits** Risk audits are specifically used to examine the implementation of response plans and their effectiveness at dealing with risks and their root causes. They are also used to determine the effectiveness of the risk management processes.

**Variance and Trend Analysis** Variance and trend analysis looks at the planned results in comparison with the actual results. Deviations found may indicate risk.

**Technical Performance Measurement** This tool and technique compares the technical accomplishments of project milestones completed during the Executing processes to the technical milestones defined in the project Planning processes. Existing variations may indicate risk.

**Reserve Analysis** Reserve analysis compares the amount of contingency reserve remaining to the amount of risk remaining in the project. The idea is to determine whether the contingency reserve is sufficient.

**Status Meetings** Status meetings are used as a means of discussing risk with the project team.

## Outputs of Monitor and Control Risks

The Monitor and Control Risks process produces the following five outputs:

- Risk register updates
- Organizational process assets updates
- Change requests
- Project management plan updates
- Project document updates

**Risk Register Updates** The risk register may need the following updates:

- Addition of newly identified risks
- Updates to the probability and impact, priority, response plans, and ownership of previously identified risks
- Outcome of the project's risks and their risk responses

**Organizational Process Assets Updates** The following updates are typically made to the organizational process assets:

- Risk management templates
- Risk breakdown structure (RBS)
- Lessons learned

**Change Requests** Recommended corrective actions or preventive actions may result from contingency plans or workarounds that were implemented as part of the Monitor and Control Risks process. Change requests are submitted to the Perform Integrated Change Control process.

**Project Management Plan Updates** The project management plan may include updates to the following documents, among others:

- Schedule management plan
- Cost management plan
- Quality management plan
- Procurement management plan
- Human resource management plan
- WBS within the scope baseline
- Schedule baseline
- Cost performance baseline

**Project Document Updates** The result of carrying out this process typically includes updates to the assumptions log and technical documentation.

**Exam Essentials**

**Describe the purpose of the Monitor and Control Risks process.** Monitor and Control Risks involves identifying and responding to new risks as they occur. Risk monitoring and reassessment should occur throughout the life of the project.

# Assessing Corrective Actions

As the need for corrective actions arise, the team may respond by submitting changes requests. All change requests are reviewed by the change control board through the Perform Integrated Change Control process. The issue register should be assessed regularly for needed corrective action so that unresolved issues receive the appropriate follow-up. Responses to open issues should be made by using the necessary tools and techniques so that impact on the schedule, cost, and resources is minimized.

## Perform Integrated Change Control

The Perform Integrated Change Control process manages the project's change requests. This process approves or rejects changes and also manages changes to the deliverables. All change requests generated by other processes are submitted to the Perform Integrated Change Control process for review before implementation. Changes may involve corrective actions, preventive actions, and defect repair.

According to the *PMBOK® Guide*, the responsibilities of the Perform Integrated Change Control process are primarily as follows:

- Influencing the factors that cause change control processes to be circumvented
- Promptly reviewing and analyzing change requests as well as reviewing and analyzing corrective and preventive actions
- Coordinating and managing changes across the project
- Maintaining the integrity of the project baselines and incorporating approved changes into the project management plan and other project documents
- Documenting requested changes and their impacts

Figure 5.15 shows the inputs, tools and techniques, and outputs of the Perform Integrated Change Control process.

**FIGURE 5.15** Perform Integrated Change Control process

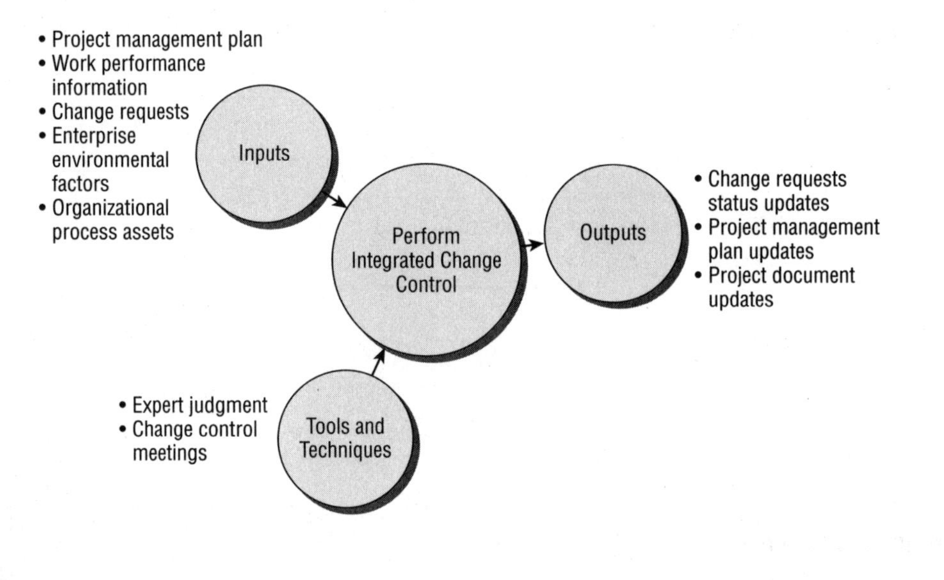

For more detailed information on the Perform Integrated Change Control process, see Chapter 10 in *PMP: Project Management Professional Exam Study Guide, 6th Edition.*

## Inputs of Perform Integrated Change Control

There are five inputs of the Perform Integrated Change Control process:

- Project management plan
- Work performance information

- Change requests
- Enterprise environmental factors
- Organizational process assets

**Project Management Plan** The project management plan includes several documents that will be utilized in this process, such as the project schedule, budget, and scope statement. These documents will be used when reviewing submitted changes.

**Work Performance Information** Work performance information includes performance results, such as status of deliverables, schedule progress, and costs incurred to date.

**Change Requests** Change requests include requests that have been submitted for review and approval, such as corrective action, preventive action, and defect repairs. All change requests must be submitted in writing. Figure 5.16 reiterates that all change requests are submitted as inputs into the Perform Integrated Change Control process.

**FIGURE 5.16** Change requests

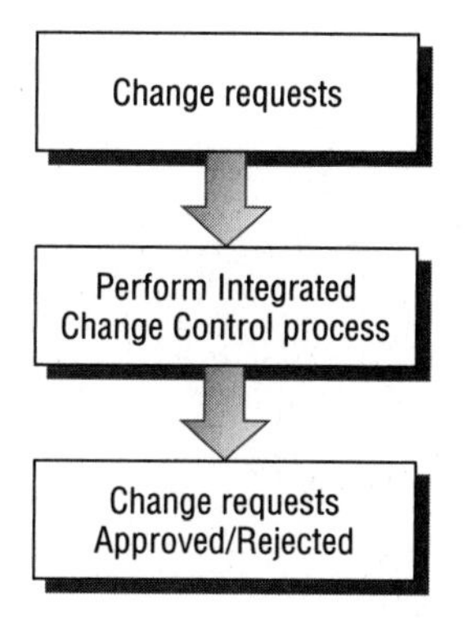


### Managing Changes

Modifications to the project are submitted in the form of change requests and managed through the Perform Integrated Change Control process. Here are some examples of factors that might cause change:

- Project constraints
- Stakeholder requests
- Team member recommendations
- Vendor issues

**Enterprise Environmental Factors** Enterprise environmental factors used to carry out this process include the PMIS, which includes the configuration management system and scheduling software.

**Organizational Process Assets** The following organizational process assets are utilized by this process:

- Change control and authorization procedures
- Process measurement database
- Configuration management knowledge base
- Project files

## Tools and Techniques of Perform Integrated Change Control

The Perform Integrated Change Control process has two tools and techniques that you will need to be familiar with.

**Expert Judgment** The following sources of expert judgment may be used in this project:

- The project management team
- Key stakeholders
- Subject matter experts
- Consultants
- Industry or professional groups
- The project management office, if applicable

**Change Control Meetings** Change control meetings are meetings in which the change control board reviews and makes decisions on submitted change requests. The change control board is established to review all change requests and is given the authority to approve or deny them. This authority is defined by the organization. The decisions should be documented and reviewed by stakeholders.

The *PMBOK® Guide* states that the change control board is made up of a group of stakeholders who review and then approve, delay, or reject change requests.

You will also need to know about configuration management systems and change control systems, which are not listed as tools and techniques of the process but directly influence how the process is carried out.

**Configuration Management System** Configuration management is generally a subsystem of the PMIS. It involves managing approved changes and project baselines. According to the *PMBOK® Guide*, the following activities are associated with configuration change management in this process:

**Configuration Identification** Configuration identification describes the characteristics of the product, service, or result of the project. This description is the basis used to verify when changes are made and how they're managed.

**Configuration Status Accounting** This includes the approved configuration identification, the status of proposed changes, and the status of changes currently being implemented.

**Configuration Verification and Auditing** Verification and audits are performed to determine whether the configuration item is accurate and correct and to make certain the performance requirements have been met.

It is the project manager's responsibility to manage changes and see that organizational policies regarding changes are implemented.

**Change Control System** Change control systems are documented procedures that describe how the deliverables of the project and associated project documentation are controlled, changed, and approved. Here are some things to note about change control systems:

- Often describe how to submit change requests and how to manage change requests
- Are usually subsystems of the configuration management system
- Track the status of change requests
- Ensure that changes that are not approved are tracked and filed in the change control log for future reference
- Define procedures that detail how emergency changes are approved

In addition, you should know about implementing these two types of systems. The *PMBOK® Guide* gives these three objectives for implementing and using configuration management systems and change control processes:

- Establish a method to consistently identify changes, request changes to project baselines, and analyze and determine the value and effectiveness of the changes.
- Continuously authenticate and improve project performance by evaluating the impact of each change.
- Communicate all change requests, whether approved, rejected, or delayed, to all the stakeholders.

Figure 5.17 provides a brief overview of the configuration management and change control systems.

**FIGURE 5.17** Configuration management system

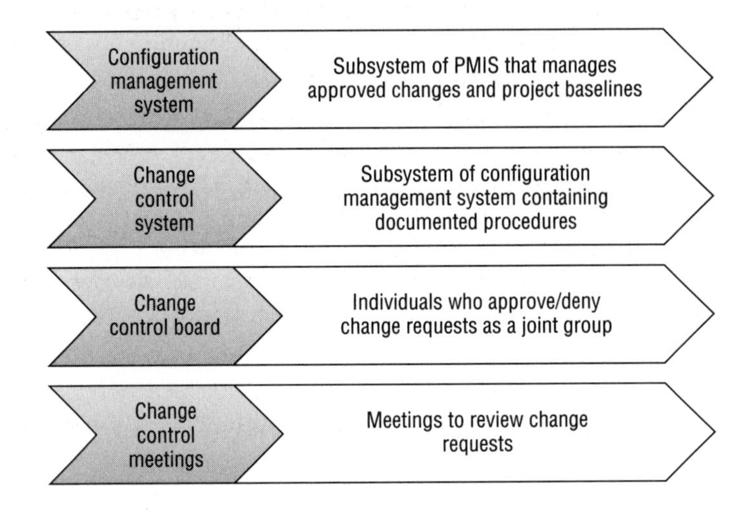

The change control system and configuration management system together identify, document, and control the changes to the performance baseline.

## Outputs of Perform Integrated Change Control

Know the three outputs of the Perform Integrated Change Control process:

**Change Requests Status Updates** Change requests status updates reflect the approval status of change requests submitted as inputs, whether approved or rejected.

**Project Management Plan Updates** Project management plan updates are made to reflect changes that have been approved or rejected after going through the Perform Integrated Change Control process. This may include changes to the following documents:

- Baselines
- Subsidiary management plans

**Project Document Updates** Project document updates may include updates to any documents impacted by approved or rejected changes as well as the change request log, which will reflect the status update of the change requests.

**Exam Essentials**

**Describe the purpose of the Perform Integrated Change Control process.** Perform Integrated Change Control is performed throughout the life of the project and involves reviewing all the project change requests, establishing a configuration management and change control process, and approving or denying changes.

**Be able to define the purpose of a configuration management system.** Configuration management systems are documented procedures that describe the process for submitting change requests, the processes for tracking changes and their disposition, the processes for defining the approval levels for approving and denying changes, and the process for authorizing the changes. Change control systems are generally a subset of the configuration management system. Configuration management also describes the characteristics of the product of the project and ensures accuracy and completeness of the description.

**Be able to describe the purpose of a change control board.** The change control board has the authority to approve or deny change requests. Its authority is defined and outlined by the organization. A change control board is made up of stakeholders.

# Communicating Project Status

Communicating project status occurs throughout activities related to the Executing and Monitoring and Controlling processes. Gathering project status from stakeholders and collecting their feedback is necessary to generating project reports. Project reports help to align the project with the business needs.

Collecting information and feedback and performing further analysis on the project's performance occur within the Report Performance process. It is in this process that project reports are generated and then handed over to the Distribute Information process for dissemination to the appropriate stakeholders.

## Report Performance

The Report Performance process collects, documents, and reports baseline data. Report Performance is part of the Project Communications Management Knowledge Area. Thus, it involves collecting actual performance data and comparing it to baseline data. This information is then reported to project team members, the management team, and other interested parties as specified in the communications management plan. Reporting might include status updates on work completed, status of risks and issues, analysis of performance, and approved changes.

Figure 5.18 shows the inputs, tools and techniques, and outputs of the Report Performance process.

**FIGURE 5.18** Report Performance process

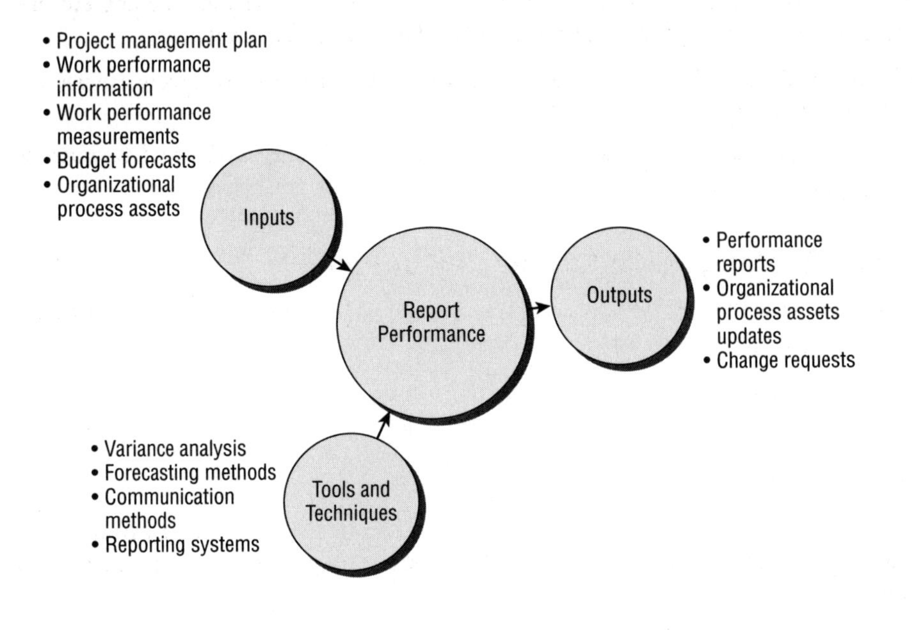

For more detailed information on the Report Performance process, see Chapter 10 in *PMP: Project Management Professional Exam Study Guide, 6th Edition.*

## Inputs of Report Performance

The Report Performance process includes the following inputs:

- Project management plan
- Work performance information
- Work performance measurements
- Budget forecasts
- Organizational process assets

**Project Management Plan** The project management plan contains the project management baseline data used to monitor and compare results. Deviations from this data are reported to management.

**Work Performance Information** Work performance measures are taken primarily during the Control Cost, Control Schedule, and Perform Quality Control processes. Information such as status of deliverables, schedule, and costs is used in the Report Performance process.

**Work Performance Measurements** Work performance measurements are used to create activity metrics that will allow for the comparison of actual to planned performance. The items evaluated for work performance include schedule, cost, and technical performance.

**Budget Forecasts** Budget forecasts look at the forecasted costs needed to complete the remaining project work and the cost forecast for the project as a whole.

**Organizational Process Assets** The following organizational process assets are used in the Report Performance process:

- The organization's policies and procedures for defining metrics and variance limits
- Report templates

## Tools and Techniques of Report Performance

You should know the following tools and techniques of the Report Performance process:

- Variance analysis
- Forecasting methods
- Communication methods
- Reporting systems

**Variance Analysis** Variance analysis determines the cause of any existing variances between baselines and actual performance. The steps used to conduct variance analysis may look something like the following:

1. Verify quality of information used.
2. Identify existing variances.
3. Determine impact of variances on the schedule and budget.
4. Analyze variance trends.

**Forecasting Methods** According to the *PMBOK® Guide*, forecasting methods fall into four categories:

**Time Series Methods** Time series methods include earned value, moving average, extrapolation, trend estimation, linear prediction, and growth curve.

**Causal/Econometric Methods** The methods included in this category are regression analysis, autoregressive moving average (ARMA), and econometrics.

**Judgmental Methods** Judgmental methods use opinions, intuitive judgments, and probability estimates to determine possible future results. Methods in this category include composite forecasts, Delphi method, surveys, technology forecasting, scenario building, and forecast by analogy.

**Other Methods** Other types of forecasting methods include simulation, probabilistic forecasting, and ensemble forecasting.

**Communication Methods** Communication methods used in this process include status review meetings. The purpose of a status meeting is to provide updated information regarding the progress of the project. Note the following about status review meetings:

- Status review meetings are a venue for formally exchanging project information.
- The project manager is usually the expediter.

**Reporting Systems** Reporting systems are used to record, store, and distribute information about the project, including cost, schedule progress, and performance information. According to the *PMBOK® Guide* the following are types of distribution formats:

- Spreadsheets and their analysis
- Presentation documents
- Tables
- Graphics

## Outputs of Report Performance

The Report Performance process results in three outputs.

**Performance Reports** Performance reports document and report performance information to the stakeholders as outlined in the communications management plan. These reports might take many forms:

- S curves (as cost performance baselines are recorded)
- Bar charts
- Tables
- Histograms
- Earned value information
- Variance analysis

According to the *PMBOK® Guide*, performance reports may range from simply stated status reports to highly detailed reports:

- Dashboards
- Analysis of project performance for previous periods
- Risk and issue status
- Work completed in the current reporting period
- Work expected to be completed during the next reporting period
- Changes approved in the current reporting
- Results of variance analysis
- Time completion forecasts and cost forecasts
- Other information stakeholders want or need to know

**Organizational Process Assets Updates** The organizational process assets may include updates to the following items:

- Lessons-learned documentation
- Report formats
- Causes of issues
- Corrective actions taken

**Change Requests** The following changes may be requested:

- Recommended corrective actions to bring performance in alignment with the project management plan
- Preventive actions to reduce probable future negative performance

**Exam Essentials**

**Describe the purpose of the Report Performance process.** Report Performance involves collecting and distributing performance information about the project, including status reports, progress to date, and forecasts.

# Bringing the Processes Together

In this chapter, you learned all about the processes that fall in the Monitoring and Controlling process group. If anything is amiss, the processes in this group work together to bring the project back in line with the project management plan.

In addition to correcting any differences between the actual work and the project management plan, the Monitoring and Controlling process group is responsible for preventive actions. During these processes, changes are verified and managed, and the status of identified risks is monitored in order to manage hurdles as soon as they arise. This process group is the project's gatekeeper.

Figure 5.19 summarizes the objectives in the Monitoring and Controlling process group.

**FIGURE 5.19** Process group objectives: Monitoring and Controlling

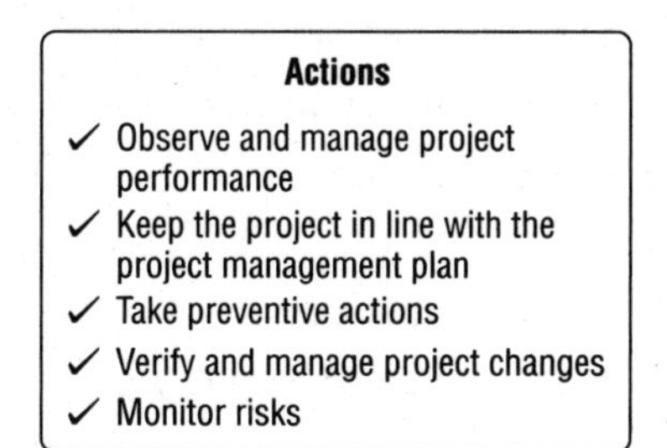

In total, the Monitoring and Controlling process group contains eight processes, covering nearly all of the knowledge areas:

- Project Integration Management
- Project Scope Management
- Project Time Management
- Project Cost Management
- Project Quality Management
- Project Communications Management
- Project Risk Management
- Project Procurement Management

We will review the process interactions that occur in each knowledge area during the Monitoring and Controlling process group. This will help you understand how the processes work together.

## Project Integration Management Knowledge Area Review

Two key processes in the Monitoring and Controlling process group belong to the Project Integration Management Knowledge Area. Figure 5.20 shows how the project work is tracked and regulated as part of this knowledge area and how all project changes are approved or rejected. Changes are reviewed by the integrated change control system, which determines whether the changes are necessary and in the best interest of the project.

**FIGURE 5.20** Process interaction: integration

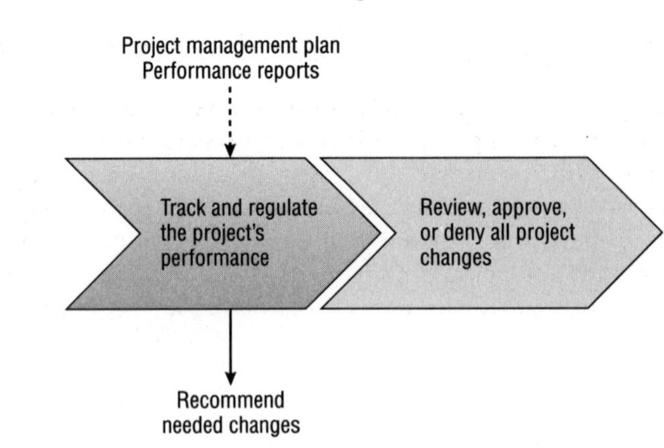

## Project Scope Management Knowledge Area Review

As you may recall, scope documents are necessary to keep the project focused by providing it with boundaries and defining exactly what it set out to accomplish (no more, no less). The two processes in the Project Scope Management Knowledge Area fall in the Monitoring and Controlling process group and ensure that the project is in line with the defined requirements.

Figure 5.21 shows how validated deliverables are formally accepted during this phase and how the project and product scope are monitored and managed. Changes made to the project scope are carefully watched. If something does not align with the project's defined requirements, then recommended changes are made. Performance measurements are also recorded and are used in project quality processes and project communications.

**FIGURE 5.21** Process interaction: scope

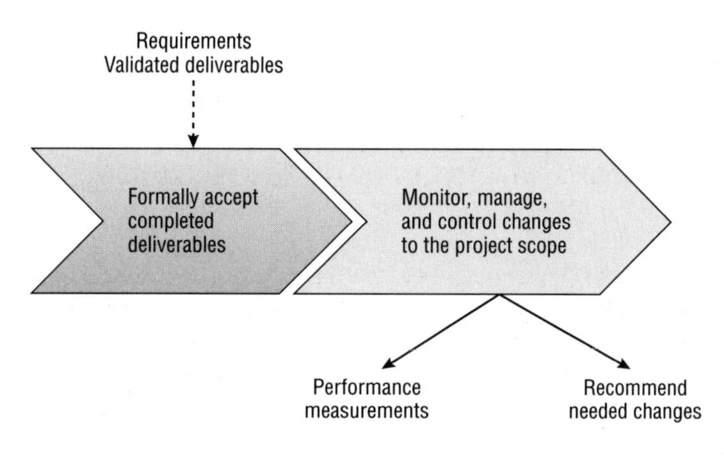


## Project Time Management Knowledge Area Review

Only one process from the Project Time Management Knowledge Area occurs in the Monitoring and Controlling process group. This process is responsible for monitoring and controlling changes to the project schedule. Here, we are also concerned with monitoring the status of the project so that progress updates can be made. Performance measurements emerge from this step. The performance measurements will be important to monitoring and controlling quality and communications.

As Figure 5.22 shows, changes are recommended based on the progress and status of the project schedule.

**FIGURE 5.22** Process interaction: time

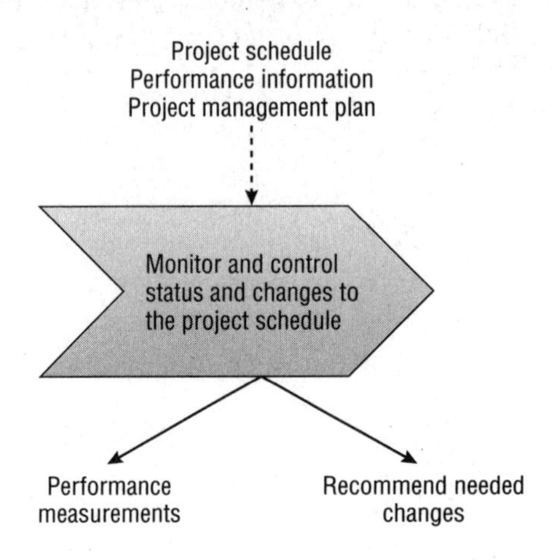


## Project Cost Management Knowledge Area Review

During the monitoring and controlling of project costs, we utilize the budget developed during the execution of the project work. Figure 5.23 shows the single cost-related process in the Monitoring and Controlling process group.

**FIGURE 5.23** Process interaction: cost

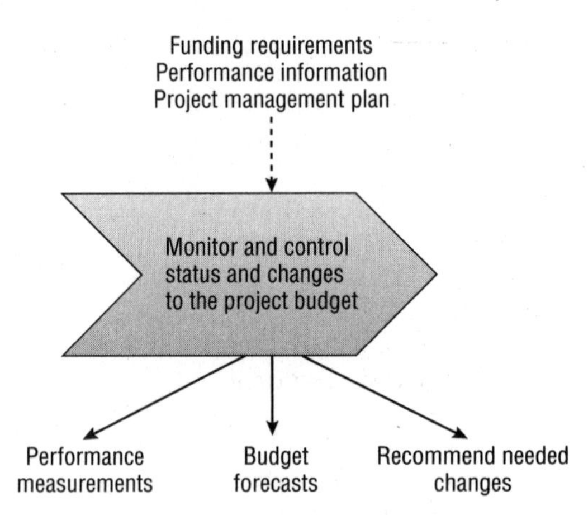


As you may recall during our review of this process earlier in this chapter, the following items are generated out of this process:

- Budget forecasts
- Performance measurements, which will be used for monitoring and controlling quality and communications (similar to the performance information of the project schedule)
- Recommended changes to bring the budget back in line with the project management plan as needed

Generating the budget forecasts and performance measurements involves the use of earned value management (EVM), which itself includes several formulas. Although we won't go through the formulas and calculations here, you'll need to remember the calculations and various performance measurements in EVM:

- Planned value (PV)
- Earned value (EV)
- Actual cost (AC)
- Schedule variance (SV)
- Cost variance (CV)
- Schedule performance index (SPI)
- Cost performance index (CPI)
- Estimate at completion (EAC)
- Estimate to complete (ETC)
- To-complete performance index (TCPI)

For a review of the cost formulas, see "Control Costs" earlier in this chapter.

## Project Quality Management Knowledge Area Review

Monitoring and controlling quality involves monitoring and recording the results of quality activities that have been executed in addition to assessing performance. Figure 5.24 shows how several items carry out this purpose.

**FIGURE 5.24** Process interaction: quality

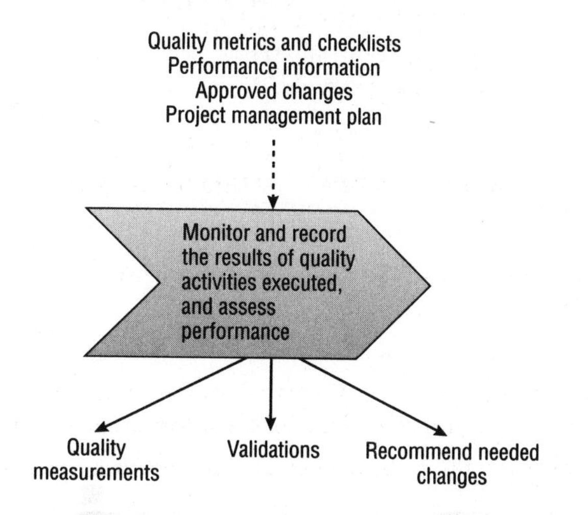

The results of the single quality-related process in the Monitoring and Controlling process group include the following items:

- Quality measurements, which are submitted back into the Perform Quality Assurance process (which is part of the Executing process group)
- Validated deliverables and changes, to make sure the results of the project work are correct
- Recommended changes, which are made for changes or deliverables that have been rejected from a quality standpoint

To generate these results, you may recall that several tools and techniques were used. You should be familiar with the quality tools for the exam. Also note that those conducting quality control activities should have a working knowledge of sampling and probability. At the least, project teams should be familiar with the following:

- Prevention
- Attribute sampling
- Tolerances and control limits

## Project Communications Management Knowledge Area Review

Monitoring and controlling project communications relies on performance information and budget forecasts; these are collected and distributed as performance reports. The recipients of these reports are outlined in the communications management plan. Figure 5.25 displays the results of the communications process in the Monitoring and Controlling process group.

**FIGURE 5.25** Process interaction: communications

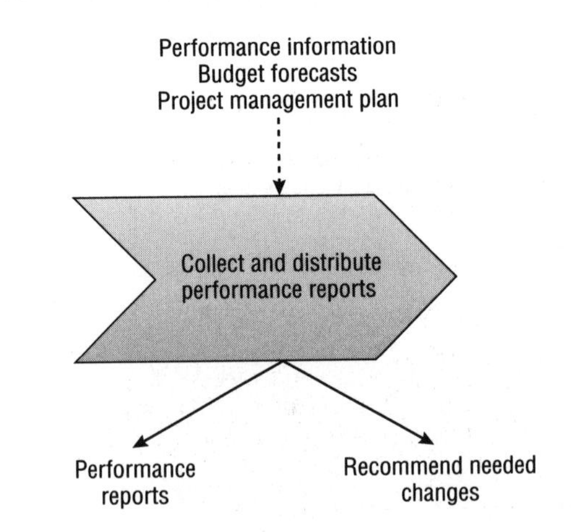


This process generates the following items:

- The performance reports, such as status reports, measurements of the project's progress, and forecasts

- Any recommended changes made to bring the project back in line with the project management plan, implement preventive actions, or accommodate changes that were made to the product or project as a result of managing stakeholder expectations

US AIR COACH
$545 SF - LOUISVILLE
STANDIFORD field
SDF

IntentMEDIA.NET

ORBITZ.COM
PRICELINE.COM
CHEAPOAIR.COM
TRAVEL200.COM

## ...rea Review

...cess that falls in the ...his single process accom- ...rry out the following

...he Planning process group

... in the following:

...y future projects

...nducted to evaluate the ...t processes

...h the project manage- ...ypically includes the

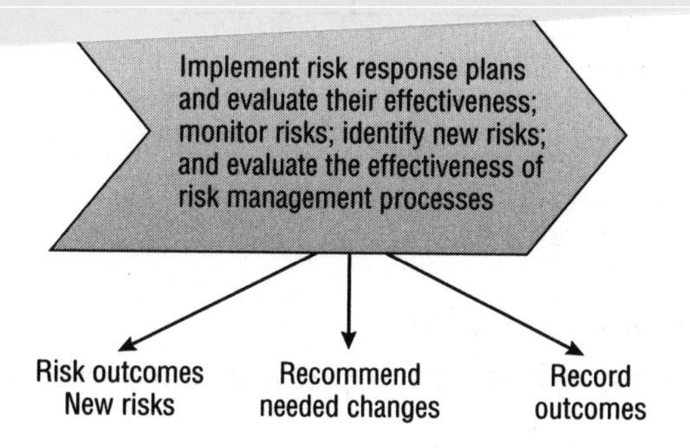

# Project Procurement Management Knowledge Area Review

During monitoring and controlling of the project, vendor relationships and performance are managed and monitored. Figure 5.27 shows what occurs in the single procurement-related process that falls in the Monitoring and Controlling process group.

**FIGURE 5.27** Process interaction: procurement

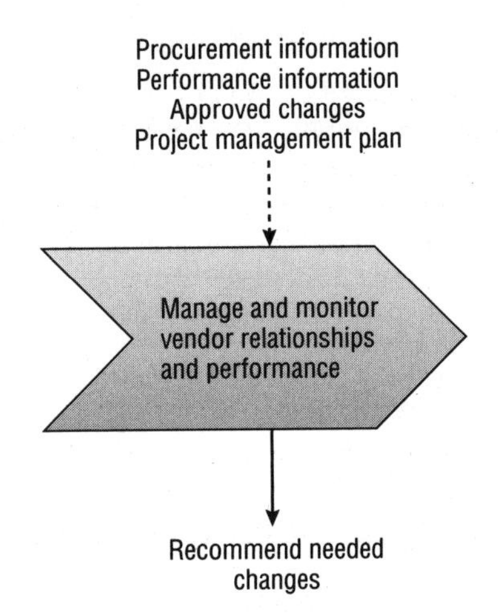


The following information is used throughout this process:

- Procurement information, such as the procurement contracts and any documents relating to the procurements being administered
- Vendor performance information, including internal performance reports and reports generated by the vendors
- Approved changes, which typically involve changes to the contract terms and conditions
- The project management plan, which contains the procurement management plan

As a result of the Administer Procurements process, recommended changes are made to bring vendor performance and progress back in line with the project management plan.

Keep in mind that the Monitoring and Controlling process group's purpose is given away by its title. The processes in this process group are concerned with making sure the project (performance and status) is moving along as planned. In the event that this is not the case, corrective actions are issued and proactive measures are taken through preventive actions.

# Review Questions

1. The Monitor and Control Project Work process results in all of the following EXCEPT:
   A. Change requests
   B. Preventive actions
   C. Defect repairs
   D. Deliverables

2. A project manager of a technical training company is working on a project that will automate the company's course registration process. Due to the expertise involved, the company hired an external vendor to create the backend development of the registration system. While monitoring the project work, the project manager discovered that the vendor was two weeks behind schedule and about to miss an important milestone. What tool and technique did the project manager use to make this determination?
   A. Contract
   B. Work performance information
   C. Performance reporting
   D. Procurement documentation

3. All of the following processes integrate with the Administer Procurements process EXCEPT:
   A. Close Procurements
   B. Direct and Manage Project Execution
   C. Report Performance
   D. Perform Integrated Change Control

4. Which of the following BEST describes the contract change control system?
   A. Meetings where the change control board reviews and makes decisions on change requests submitted
   B. A subsystem of the PMIS that manages approved changes and project baselines
   C. A system that describes the processes needed to make contract changes
   D. A system that describes how the deliverables of the project and associated project documentation are controlled, changed, and approved

5. While conducting earned value management, the project manager of a software development company receives a request from a stakeholder for the most current cost estimate report. Specifically, the stakeholder needs to know the value of the work completed against actual cost. Which formula will the project manager use to calculate this information?
   A. $EV \div AC$
   B. $EV - AC$
   C. $EV - PV$
   D. $EV \div PV$

6. During a project status meeting, the project manager provided the room of stakeholders with results of a recent quality audit. After the group discussed the results, it was determined that a minor adjustment needed to be made to bring the variance levels under control. The project sponsor, also a meeting participant, agreed and gave her approval to implement the changes immediately. What should the project manager do next?

   A. Inform the quality management team and implement the changes that day.

   B. Obtain the customer's approval for the change, even though the project sponsor has approved it.

   C. Document the change request and submit it for review by the change control board.

   D. Bypass the change control process because the change has already received approval by a high authority.

7. After the implementation of yet another risk response plan, the project manager became concerned about whether the budget was sufficient enough to cover the remaining identified risks. Which of the following tools and techniques can the project manager use to evaluate the status of the budget in terms of risk?

   A. Risk reassessment

   B. Risk audits

   C. Budget analysis

   D. Reserve analysis

8. If budget at completion (BAC) is 1,500 and estimate at completion (EAC) is 1,350, what is the current variance at completion (VAC)?

   A. 150

   B. 75

   C. 0.90

   D. 1.11

9. A project manager is in the process of determining the existing variance between the schedule baseline and actual schedule. To determine the variance, the project manager used schedule variance (SV) and schedule performance index (SPI). The project manager is currently in what process?

   A. Control Costs

   B. Control Schedule

   C. Perform Quality Control

   D. Report Performance

10. What level of accuracy does 3 sigma represent?

    A. 68.26 percent

    B. 95.46 percent

    C. 99.73 percent

    D. 99.99 percent

# Answers to Review Questions

1. D. Deliverables are an output of the Direct and Manage Project Execution process, not of the Monitor and Control Project Work process. Preventive actions and defect repairs (options B and C) are part of change requests.

2. C. Options A and B are inputs of the Administer Procurements process, and option D is an output. This leaves C, performance reporting. Performance reporting provides information about the vendor's progress in meeting contract objectives. This question simply tested your ability to distinguish between the inputs, tools and techniques, and outputs of the process.

3. A. According to the *PMBOK® Guide*, you must integrate and coordinate the Direct and Manage Project Execution, Report Performance, Perform Quality Control, Perform Integrated Change Control, and Monitor and Control Risk processes during the Administer Procurements process. Only Close Procurements, option A, is not included in this list of processes.

4. C. Option A refers to change control meetings, option B refers to the configuration management system, and option D refers to the change control system. Also notice that C, the correct choice, is the only option that specifically mentions contracts.

5. A. In this scenario, the stakeholder is looking for the latest cost performance index, which measures the value of the work completed against actual cost. The correct formula simply takes earned value and divides it by actual cost.

6. C. Change requests should always be documented, even if verbal approval has been provided by a high authority figure of the project (such as the project sponsor in this scenario). Obtaining the approval of the project sponsor may speed up the approval process, but it may not replace it.

7. D. You may have been tempted to choose option C. However, budget analysis is not a tool and technique of the Monitor and Control Risks process, which is the process used in this scenario. The correct choice is reserve analysis. Reserve analysis compares the amount of unused contingency reserve to the remaining amount of risk. This information is then used to determine whether a sufficient amount of reserve remains to deal with the remaining risks.

8. A. To calculate variance at completion, simply subtract the estimate at completion (EAC) from the budget at completion (BAC). Plug in the numbers provided in the question and work out the problem as follows:

   $$VAC = 1{,}500 - 1{,}350$$

**9.** B. The project manager is currently in the Control Schedule process. This can be a tricky question because PV and SPI were mentioned, steering you toward the Control Costs process. In this scenario, the project manager is specifically utilizing the variance analysis technique of the Control Schedule process.

**10.** C. For reference, know the level of quality for 1 sigma (68.26 percent level of accuracy), 2 sigma (95.46 percent level of accuracy), 3 sigma (99.73 percent level of accuracy), and 6 sigma (99.99 percent level of accuracy).

Chapter 6

# Closing the Project

## THE PMP EXAM CONTENT FROM THE CLOSING THE PROJECT PERFORMANCE DOMAIN COVERED IN THIS CHAPTER INCLUDES THE FOLLOWING:

- ✓ Obtain final acceptance of the project deliverables by working with the sponsor and/or customer, in order to confirm that project scope and deliverables were met.
- ✓ Transfer the ownership of deliverables to the assigned stakeholders in accordance with the project plan, in order to facilitate project closure.
- ✓ Obtain financial, legal, and administrative closure using generally accepted practices, in order to communicate formal project closure and ensure no further liability.
- ✓ Distribute the final project report including all project closure-related information, project variances, and any issues, in order to provide the final project status to all stakeholders.
- ✓ Collate lessons learned through comprehensive project review, in order to create and/or update the organization's knowledge base.
- ✓ Archive project documents and materials in order to retain organizational knowledge, comply with statutory requirements, and ensure availability of data for potential use in future projects and internal/external audits.
- ✓ Measure customer satisfaction at the end of the project by capturing customer feedback, in order to assist in project evaluation and enhance customer relationships.

Closing is the final process group of the five project management process groups and accounts for 8 percent of the questions on the PMP exam. The primary purpose of Closing is to formally complete the project, phase, or contractual obligations. Finalizing all of the activities across the project management process groups accomplishes this.

By the end of the processes that make up the Closing process group, the completion of the defined processes across all of the process groups will have been verified.

The process names, inputs, tools and techniques, outputs, and descriptions of the project management process groups and related materials and figures in this chapter are based on content from *A Guide to the Project Management Body of Knowledge, 4th Edition (PMBOK® Guide).*

# Obtaining Final Acceptance

All projects must be formally closed out regardless of the reasons for closure (such as successful completion or early termination). According to the *PMBOK® Guide*, the following typically occurs during formal closing of the project or project phase:

- Obtaining formal acceptance by the customer or sponsor
- Closing out procurement contracts
- Conducting phase-end or post-project reviews
- Updating the organizational process assets, which includes documenting lessons learned and archiving project documents in the project management information system (PMIS)
- Recording the impacts of tailoring to any process

These activities are carried out through the Close Project or Phase process and the Close Procurements process.

## Close Project or Phase

The Close Project or Phase process is concerned with gathering project records and disseminating information to formalize the acceptance of the product, service, or result. The process belongs to the Project Integration Management Knowledge Area. Here are additional notes about the process you should know:

- It involves analyzing the project management processes to determine their effectiveness.
- It documents lessons learned concerning the project processes.
- It archives all project documents for historical reference. You can probably guess that Close Project or Phase belongs to the Project Integration Management Knowledge Area since this process touches so many areas of the project.

According to the *PMBOK® Guide*, every project requires closure, and the completion of each project phase requires project closure as well. The Close Project or Phase process is performed at the close of each project phase and at the close of the project.

Figure 6.1 shows the inputs, tools and techniques, and outputs of the Close Project or Phase process.

**FIGURE 6.1** Close Project or Phase process

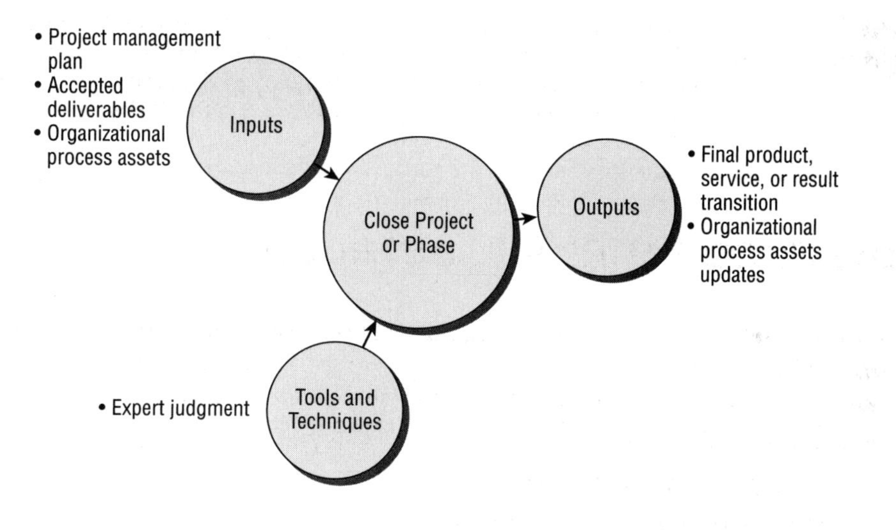

For more detailed information on the Close Project or Phase process, see Chapter 12, "Applying Professional Responsibility," in *PMP: Project Management Professional Exam Study Guide, 6th Edition* (Sybex, 2011).

## Inputs of Close Project or Phase

For the exam, know the three inputs of the Close Project or Phase process.

**Project Management Plan** To consider the project closed, the project management plan will be reviewed by the project manager to ensure completion of the deliverables.

**Accepted Deliverables** Accepted deliverables are an output of the Verify Scope process. This input is necessary to determine whether all deliverables have been successfully completed.

**Organizational Process Assets** The following organizational process assets are typically utilized within this process:

- Guidelines for closing out the project or project phase
- Historical information and lessons learned
- Project closure reports updated as part of the Plan Communications process

### Tools and Techniques of Close Project or Phase

The Close Project or Phase process has only one tool and technique: expert judgment.

Subject matter experts can help assure that the process is performed according to the organization's standard and to project management standards.

### Outputs of Close Project or Phase

There are two outputs that result from carrying out the Close Project or Phase process.

**Final Product, Service, or Result Transition** Final product, service, or result transition refers to the acceptance of the product and the turnover to the customer or into operations. This is where information is distributed that formalizes project completion.

**Organizational Process Assets Updates** The following organizational process assets are typically updated as a result of the Close Project or Phase process:

- Project files, such as the project planning documents, change records and logs, and issue logs
- Project or phase closure documents, which include documentation showing that the project or phase is completed and that the transfer of the product of the project to the organization has occurred
- Project closure reports
- Historical information, which is used to document the successes and failures of the project

## Close Procurements

The Close Procurements process is primarily concerned with completing and settling the terms of the procurement. It supports the Close Project or Phase process by determining whether the work described in the procurement documentation or contract was completed accurately and satisfactorily, which is referred to as product verification.

The Close Procurements process also accomplishes the following:

- Updating and archiving records for future reference
- Completing all terms or conditions for completion and closeout specified within the procurement contracts
- Documenting and verifying project outcomes

Figure 6.2 shows the inputs, tools and techniques, and outputs of the Close Procurements process.

**FIGURE 6.2** Close Procurements process

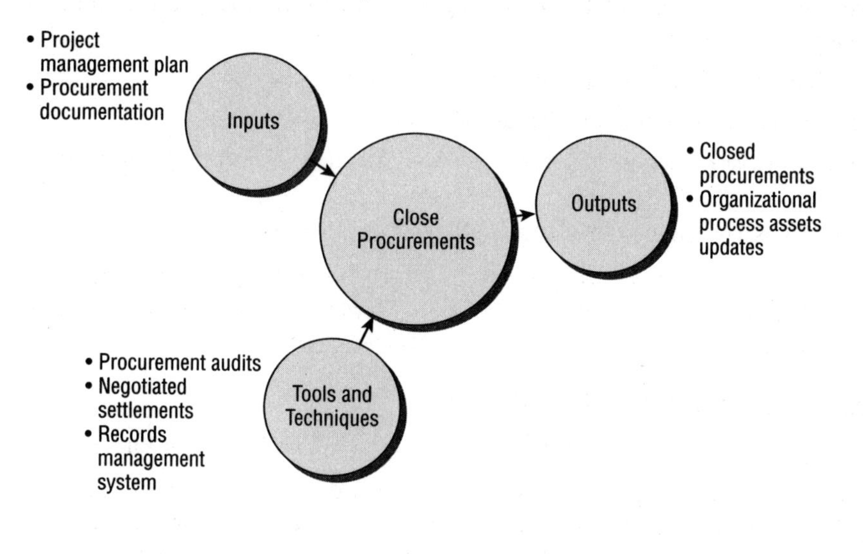

For more detailed information on the Close Procurements process, see Chapter 12 of the *PMP: Project Management Professional Exam Study Guide, 6th Edition.*

## Inputs of Close Procurements

The Close Procurements process has a total of two inputs that you should be familiar with: project management plan and procurement documentation.

**Project Management Plan** The project management plan contains the procurement management plan, which will be used to carry out this process.

**Procurement Documentation** According to the *PMBOK® Guide*, the following documents from within the procurement documentation input are collected, indexed, and filed:

- Contract schedule
- Scope
- Quality
- Cost performance
- Contract change documentation
- Payment records
- Inspection results

## Tools and Techniques of Close Procurements

For the exam, be familiar with the three tools and techniques of the Close Procurements process.

**Procurement Audits** Procurement audits are structured reviews that span all of the procurement processes. The primary purpose of these audits is to identify lessons learned by examining the procurement processes to determine areas of improvement and to identify flawed processes or procedures.

**Negotiated Settlements** Negotiated settlements occur when using an alternative dispute resolution (ADR) technique due to outstanding issues or claims. Reaching a negotiated settlement through the aid of ADR techniques is the most favorable way to resolve a dispute.

**Records Management System** The records management system contains processes, control functions, and automation tools to help manage contract and procurement documentation and records.

## Outputs of Close Procurements

You should know the two outputs of the Close Procurements process, which are closed procurements and organizational process assets updates.

**Closed Procurements** Closed procurements are the formal acceptance and closure of the procurements. As Figure 6.3 illustrates, this output satisfies the terms and conditions specified within the contract; it is also included within the procurement management plan.

**FIGURE 6.3** Closed procurements

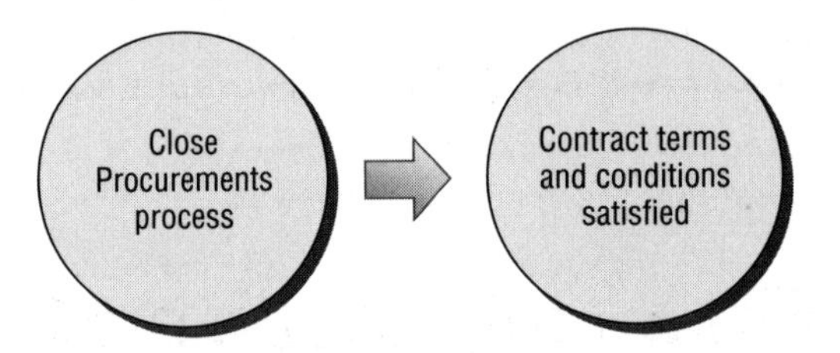


**Organizational Process Assets Updates** Updates to the organizational process assets, as a result of carrying out this process, are typically made to the following items:

- Procurement file, which includes all of the procurement records and supporting documents
- Deliverable acceptance, which is a formal written notice from the buyer that the deliverables are acceptable and satisfactory or have been rejected
- Lessons learned documentation

### Exam Essentials

**Be able to name the primary activity of the Closing processes.** The primary activity of the Closing processes is to distribute information that formalizes project completion.

**Be able to describe when the Close Project or Phase process is performed.** Close Project or Phase is performed at the close of each project phase and at the close of the project.

**Name the primary purpose of the Close Procurements process.** The Close Procurements process is primarily concerned with completing and settling the terms of the procurements.

# Transferring Ownership

The acceptance and turnover of the product to the customer or into operations typically involves a formal sign-off indicating that those signing accept the product of the project. This occurs as a result of carrying out the Close Project or Phase process, which was discussed earlier within this chapter.

### Exam Essentials

**Be able to name the process responsible for transferring ownership of the project's final product, service, or result.** The Close Project or Phase process is responsible for transferring the final product, service, or result of the project to the customer or transitioning it into operation.

# Obtaining Financial, Legal, and Administrative Closure

Aside from the two closing processes, there are a few additional items to note about the Closing process group and finalizing the project's closure. To start, project or phase closure is carried out in a formal manner and requires that financial, legal, and administrative closure be obtained. This is important to communicating formal closure and to ensuring that no further liability exists after the project has been completed.

It's important to note that a project can move into the Closing process group for several reasons:

- It's completed successfully.
- It's canceled or killed prior to completion.
- It evolves into ongoing operations and no longer exists as a project.

## Formal Project Endings

Aside from these common reasons that result in a project moving into the Closing processes, there are four formal types of project endings that you should be familiar with: addition, starvation, integration, and extinction.

**Addition** A project that evolves into ongoing operations is considered a project that ends because of addition, moving into its own ongoing business unit. Once it experiences this transition, it no longer meets the definition of a project.

**Starvation** Starvation occurs when resources are cut off from a project prior to the completion of all the requirements. This results in an unfinished project. Starvation often occurs as a result of shifting priorities, a customer's cancellation of an order or request, the project budget being reduced, or key resources quitting.

**Integration** Integration occurs when the resources of a project (such as people, equipment, property, and supplies) are distributed to other areas in the organization or are assigned to other projects. In other words, resources have been reassigned or redeployed, causing an end to the project.

**Extinction** Extinction occurs when a project has completed and stakeholders have accepted the end result. This is the best type of ending because the project team has completed what they set out to achieve.

In all cases, it's important to retain good documentation that describes why a project ended early. Performing a project review in these cases is important to retaining key details and specifics regarding why a project ended before all requirements were completed.

## Trends

Another notable piece of information involves stakeholder influence and cost trends that occur during project closure. Stakeholders tend to have the least amount of influence during the Closing processes, while project managers have the greatest amount of influence. Costs are significantly lower during the Closing processes because the majority of the project work and spending have already occurred.

## Administrative Closure

As part of administrative closure, resources will need to be released and the administrative closure procedures carried out.

### Release Resources

Although releasing project team members is not an official process, you will release your project team members at the conclusion of the project, and they will go back to their functional managers or be assigned to a new project if you're working in a matrix-type organization. The release of project resources is addressed within the staff release plan.

The staff release plan is one of the items within the staffing management plan, which in turn is included within the human resource plan, a subsidiary project management plan. The following are the primary items covered in the staff release plan, which includes the release of resources not just at the end of the project, but in general at any point within the project:

- Outlines the method for resource release
- Defines when resources are to be released

Once resources are released, the costs of the resources are no longer charged to the project. The staff release plan helps mitigate human resource risks that may occur during or at the conclusion of the project.

### Perform Administrative Closure

Administrative closure procedures involve the following:

- Collecting all the records associated with the project
- Analyzing the project success (or failure)
- Documenting and gathering lessons learned
- Properly archiving project records
- Documenting the project team members' and stakeholders' roles and responsibilities:
    - Approval requirements of the stakeholders for project deliverables and changes to deliverables
    - Confirmation that the project meets the requirements of the stakeholders, customers, and sponsor
    - Documented actions to verify that the deliverables have been accepted and exit criteria have been met

Figure 6.4 shows the elements of the administrative closure procedures.

**FIGURE 6.4** Administrative closure procedures

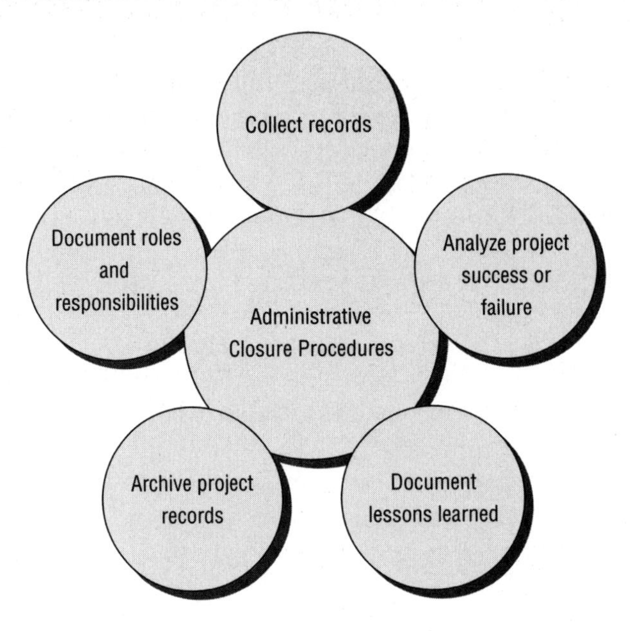


## Exam Essentials

**Be familiar with key trends projects experience during project closure.** Notable trends during project closure are that stakeholders have the least amount of influence while project managers have the greatest amount of influence and that costs are significantly lower.

**Be able to name the four formal types of project closure.** The four formal types of project closure are addition, starvation, integration, and extinction.

**Know when and how resources are released from the project.** Project resources are primarily released as part of finalizing project closure. The release of project resources is addressed in the staff release plan.

**Be familiar with the purpose of the administrative closure procedures.** Administrative closure involves collecting all the records associated with the project, analyzing the project success or failure, documenting and gathering lessons learned, properly archiving project records, and documenting roles and responsibilities.

# Distributing the Final Project Report

As part of formally closing out a project, the project manager will need to create and distribute the final project report. This generally occurs as part of the Close Project or Phase process, where administrative details are addressed. The final project report should include all project closure–related details, project variances, and any issues. The project manager should ensure that final project status is distributed to the stakeholders.

**Exam Essentials**

**Be familiar with the information distributed as part of the final project report.** The final project report distributed by the project manager to stakeholders includes project closure–related information, project variances, and issues.

# Collating Lessons Learned

One of the recurring themes communicated across the *PMBOK® Guide* is the proactive nature carried out by the project manager, expressed by their actions. An example of being proactive is documenting lessons learned, which becomes part of the organizational process assets.

Documenting lessons learned, as mentioned previously, focuses on capturing what went well within the project, what didn't go well, and what can be improved. Although lessons learned are recorded throughout the life of the project, holding a final review with the project team before the team is released at the end of the project is essential. This information can be captured through a comprehensive project review with stakeholders and then added to the organization's knowledge base.

Documenting lessons learned occurs as part of the Close Project or Phase process, through the organizational process assets updates output.

**Exam Essentials**

**Be familiar with how lessons learned are captured at the end of the project and what information is recorded.** At the end of the project, lessons learned can be documented by holding a comprehensive project review with stakeholders. The focus of lessons learned includes documenting what went well within the project, what didn't go well, and what can be improved.

# Archiving Project Documents

As part of the Close Project or Phase process, project documents are archived. This occurs through the organizational process assets updates output, which makes the historical documents available for future reference and use. Archiving project documents provides the following benefits:

- Retaining organizational knowledge
- Ensuring that statutory requirements, if applicable, are adhered to
- Making project data available for use in future projects
- Making project data available for internal/external audits

**Exam Essentials**

**Know the process and output responsible for archiving project documents.** Final project documents are archived as part of the organizational process assets updates output through the Close Project or Phase process.

# Measuring Customer Satisfaction

According to the *PMBOK® Guide*, project success is measured by the following criteria:

- Product and project quality
- Timeliness
- Budget compliance
- Degree of customer satisfaction

Customer satisfaction is an important goal you're striving for in any project. If your customer is satisfied, it means you've met their expectations and delivered the product or service as defined within the planning processes. Customer satisfaction can be measured through quality management, which in part is concerned with making sure that customer requirements are met. This is done through understanding, evaluating, defining, and managing customer and stakeholder expectations.

As Figure 6.5 illustrates, modern quality management achieves customer satisfaction in part by ensuring that the project accomplishes what it set out to do (conformance to requirements) and that it satisfies real needs (fitness for use).

**FIGURE 6.5** Measuring customer satisfaction

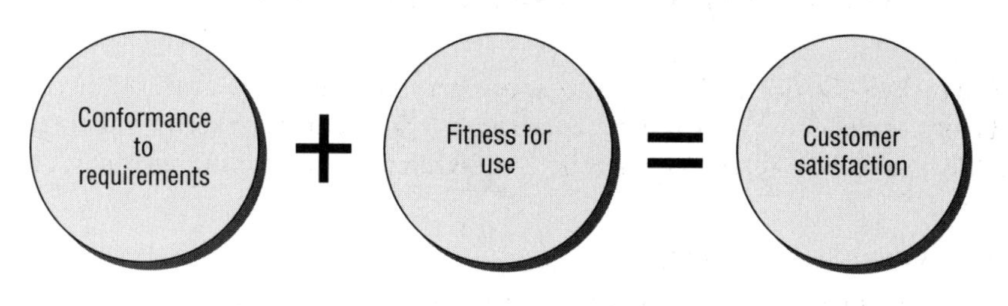


**Exam Essentials**

**Understand how to measure customer satisfaction.** Customer satisfaction is measured through quality management, which ensures that the project meets conformance to requirements and fitness for use.

# Bringing the Processes Together

The Closing process group consists of only two processes that work toward closing out the project or project phase. Before closing out the project, the project manager must first make sure that the project has met its objectives and accomplished what it set out to achieve. The Closing process group is also concerned with the following objectives:

- Obtaining formal acceptance of the project's completion
- Closing out procurement contracts
- Releasing the project resources
- Measuring customer satisfaction
- Archiving the project information for later use

Figure 6.6 reflects the objectives of the Closing process group through several key questions that the project manager should ask before considering the project fully closed.

Next, we'll go through the two closing processes and review the interactions that occur within them.

**FIGURE 6.6** Project closure checklist

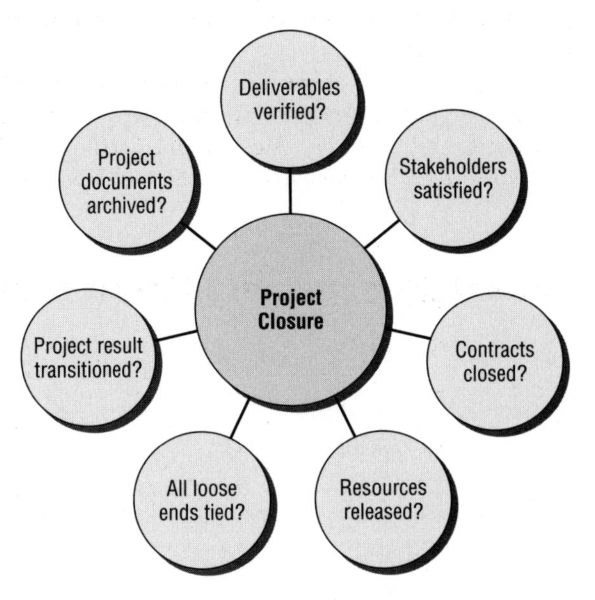


## Project Integration Management Knowledge Area Review

A project closeout cannot occur without first finalizing all project or phase activities. The Close Project or Phase process from within the Project Integration Management Knowledge Area is responsible for making this happen. Figure 6.7 shows what occurs within this process. The accepted deliverables are used to verify that the project or the project phase has accomplished what it needed to.

**FIGURE 6.7** Process interaction: integration

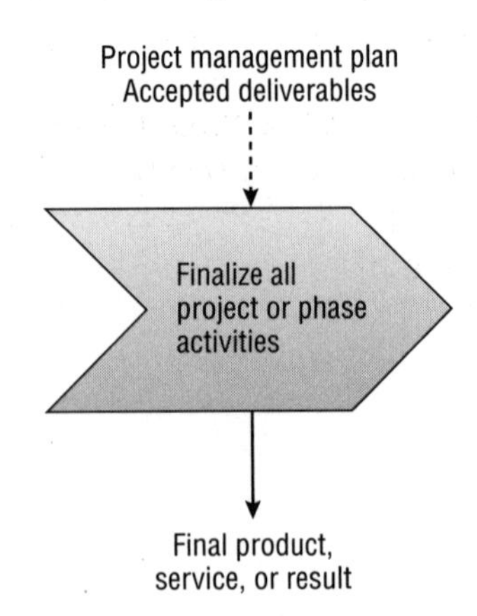

The Close Project or Phase process is also responsible for administrative closure of the project or project phase. The result is the transition to the customer of the final product, service, or result that the project created.

Before a project can be fully closed, all of the project documents must be archived within the organizational process assets. As you may recall, this includes all of the project files, any closing documents, and historical information. The historical information is archived within the lessons learned knowledge base, which will be a valuable asset to future projects.

## Project Procurement Management Knowledge Area Review

All procurement contracts must be completed and closed out before the project itself can close. This means that any loose ends or work on the vendor's end must be finalized as part of closing activities. This occurs through the Close Procurements process, which is the only closing-related process within the Project Procurement Management Knowledge Area. As Figure 6.8 illustrates, information from the project management plan and the procurement documents are used in closing out the procurement contracts.

**FIGURE 6.8** Process interaction: procurement

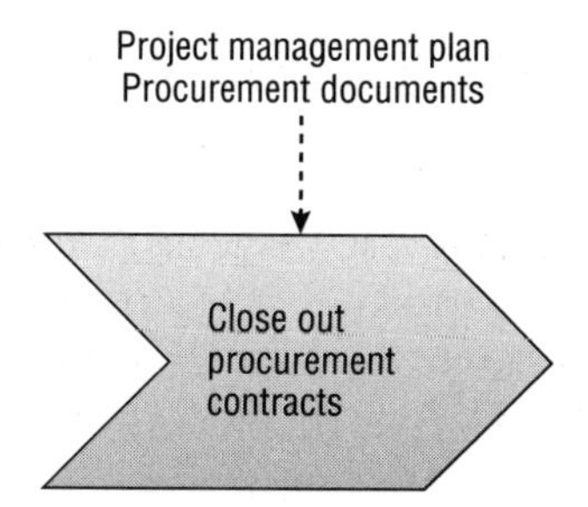


Keep in mind that procurement contracts may terminate without completion of the work. Remember that this is a special circumstance and all parties must mutually agree upon early termination.

You may have already noticed that the Close Project or Phase and the Close Procurements processes work together. Both processes confirm that the work has been completed and verified. You'll also notice that the Close Project or Phase process cannot end until the Close Procurements process has been completed.

# Review Questions

1. All of the following are inputs of the Close Project or Phase process EXCEPT:
   - **A.** Procurement documentation
   - **B.** Project management plan
   - **C.** Accepted deliverables
   - **D.** Organizational process assets

2. A project manager of a travel excursion company is currently working on confirming that all project deliverables have been successfully completed to transfer the project to the customer. What process is the project currently in?
   - **A.** Verify Scope
   - **B.** Scope Closure
   - **C.** Close Project or Phase
   - **D.** Close Procurements

3. When closing out a project or project phase, which of the following tools and techniques can be utilized to ensure that closure has been performed to the appropriate standards?
   - **A.** Records management system
   - **B.** Procurement audits
   - **C.** Expert judgment
   - **D.** Organizational process assets

4. The Close Procurements process accomplishes all of the following EXCEPT:
   - **A.** Updating and archiving records for future reference
   - **B.** Completing closeout terms or conditions as specified within the procurement contracts
   - **C.** Documenting and verifying project outcomes
   - **D.** Transitioning the final product, service, or result to the organization

5. While conducting project closure activities, a relatively new project manager of a white board production company becomes confused as to how the organization verifies that the existing criteria have been met. The project manager has found that some of the procedures followed by the organization are different from previous companies she's worked for. Where can the project manager go to clarify the procedures for verifying that the exit criteria have been met?
   - **A.** Administrative closure procedures
   - **B.** Close Project or Phase
   - **C.** Close Procurements
   - **D.** Expert judgment

6. All of the following are tools and techniques of the Close Procurements process EXCEPT:
   A. Procurement documentation
   B. Procurement audits
   C. Negotiated settlements
   D. Record management system

7. Procurement audits are concerned with which of the following?
   A. Resolving issues
   B. Documenting the processes, control functions, and automation tools
   C. Identifying flawed scope processes
   D. Identifying lessons learned

8. A project manager was in the midst of trying to resolve a dispute with one of the project's largest vendors. The issue revolved around the responsibility of the vendor to comply with changes to the scope. Although this was addressed in the contract, the vendor felt that the scope changes were much larger than the scope of the contract itself. What would be the BEST way to resolve the dispute?
   A. Through the court system
   B. Using an alternative dispute resolution technique
   C. Terminating the contract
   D. Making changes to the contract

9. Project managers have the greatest amount of influence during which stage of the project?
   A. Prior to the start of the project
   B. During the Planning processes
   C. During execution of the project work
   D. During the Closing processes

10. Resource release criteria can BEST be found in which plan?
    A. The project management plan
    B. The staff release plan
    C. The staffing management plan
    D. The human resource plan

# Answers to Review Questions

1. A. Procurement documentation is an input utilized within the Close Procurements process, the second and only other process within the Closing process group. Options B, C, and D are the only three inputs of the Close Project or Phase process.

2. C. You may have leaned toward selecting option A at first glance because deliverables are verified during the Verify Scope process. However, the project manager is currently confirming that "all deliverables" of the project have been successfully accomplished, not verifying the actual deliverables themselves. To make this determination, the Close Project or Phase process utilizes the project management plan and the list of accepted deliverables.

3. C. This question refers to the Close Project or Phase process, which has only one tool and technique: expert judgment. This is used when performing administrative closure. Options A and B are tools and techniques of the Close Procurements process, and option D is an input to the Close Project or Phase process.

4. D. Transferring the final product, service, or result is an activity and output of the Close Project or Phase process.

5. A. In this scenario, the project manager would refer to the administrative closure procedures. These procedures address several closure items, such as collecting project records, analyzing the success of the project, and documenting the roles and responsibilities of the project team members and stakeholders. As part of the latter, the administrative closure procedures document the actions to verify that the project deliverables have been accepted and that the exit criteria have been met.

6. A. Procurement documentation is an input to the Close Procurements process.

7. D. Procurement audits are structured reviews spanning all of the procurement processes. These audits are concerned with identifying lessons learned and identifying flawed processes or procedures. As a side note, option A refers to negotiated settlements, option B refers to the record management system, and option C does not refer to a specific audit process.

8. B. Disputes or outstanding claims are resolved by using alternative dispute resolution techniques, which is considered to be the most favorable way of resolving disputes. Option A, taking the dispute to court, is the least favorable resolution technique.

9. D. The project manager has the greatest amount of influence during the Closing processes, while stakeholders have the least amount of influence during this stage of the project.

10. B. This is a tricky question because all of the options are technically correct. However, the question asks for the best answer. This would be option B, the staff release plan. The staff release plan is part of the staffing management plan, which is part of the human resource plan, which is part of the project management plan.

Appendix

# About the Companion CD

## IN THIS APPENDIX:

- What you'll find on the CD
- System requirements
- Using the CD
- Troubleshooting

# What You'll Find on the CD

The following sections are arranged by category and summarize the software and other goodies you'll find on the CD. If you need help with installing the items provided on the CD, refer to the installation instructions in the section "Using the CD" later in this appendix.

## Sybex Test Engine

The CD contains the Sybex test engine, which includes two bonus practice exams located only on the CD.

## Electronic Flashcards

These handy electronic flashcards are just what they sound like. One side contains a question (standard question and answer or fill-in-the-blank question), and the other side shows the answer.

## PDF of Glossary of Terms

We have included an electronic version of the glossary in PDF format. You can view the electronic version of the glossary with Adobe Reader.

## Adobe Reader

We've also included a copy of Adobe Reader so you can view PDF files that accompany the book's content. For more information on Adobe Reader or to check for a newer version, visit Adobe's website at `www.adobe.com/products/reader/`.

# System Requirements

Make sure your computer meets the minimum system requirements shown in the following list. If your computer doesn't match up to most of these requirements, you may have problems using the software and files on the companion CD. For the latest and greatest information, please refer to the ReadMe file located at the root of the CD-ROM.

- A PC running Microsoft Windows 98, Windows 2000, Windows NT4 (with SP4 or later), Windows Me, Windows XP, Windows Vista, or Windows 7
- An Internet connection
- A CD-ROM drive

# Using the CD

To install the items from the CD to your hard drive, follow these steps:

1. Insert the CD into your computer's CD-ROM drive. The license agreement appears.

NOTE *Windows users*: The interface won't launch if you have autorun disabled. In that case, click Start ➢ Run (for Windows Vista or Windows 7, Start ➢ All Programs ➢ Accessories ➢ Run). In the dialog box that appears, type **D:\Start.exe**. (Replace *D* with the proper letter if a different letter is used for your CD drive. If you don't know the letter, see how your CD drive is listed under My Computer.) Click OK.

2. Read the license agreement, and then click the Accept button if you want to use the CD.

The CD interface appears. The interface allows you to access the content with just one or two clicks.

# Troubleshooting

Wiley has attempted to provide programs that work on most computers with the minimum system requirements. Alas, your computer may differ, and some programs may not work properly for some reason.

The two likeliest problems are that you don't have enough memory (RAM) for the programs you want to use or you have other programs running that are affecting how a program installs or runs. If you get an error message such as "Not enough memory" or "Setup cannot continue," try one or more of the following suggestions and then try using the software again:

**Turn off any antivirus software running on your computer.** Installation programs sometimes mimic virus activity and may make your computer incorrectly believe that it's being infected by a virus.

**Close all running programs.** The more programs you have running, the less memory is available to other programs. Installation programs typically update files and programs, so if you keep other programs running, installation may not work properly.

**Have your local computer store add more RAM to your computer.** This is, admittedly, a drastic and somewhat expensive step. However, adding more memory can really help the speed of your computer and allow more programs to run at the same time.

## Customer Care

If you have trouble with the book's companion CD-ROM, please call the Wiley Product Technical Support phone number at (800) 762-2974. Outside the United States, call +1(317) 572-3994. You can also contact Wiley Product Technical Support at `http://sybex.custhelp.com`. John Wiley & Sons will provide technical support only for installation and other general quality-control items. For technical support on the applications themselves, consult their vendor or author.

To place additional orders or to request information about other Wiley products, please call (877) 762-2974.

# Index

**Note to the Reader:** Throughout this index, **boldfaced** page numbers indicate primary discussions of a topic. *Italicized* page numbers indicate illustrations.

## A



## B

## D

# E

## F



## G

# H



# I

## P

## Q



## R

# T

## U



## V



## W



## Z

## John Wiley & Sons, Inc. End-User License Agreement

**READ THIS.** You should carefully read these terms and conditions before opening the software packet(s) included with this book "Book". This is a license agreement "Agreement" between you and John Wiley & Sons, Inc. "JWS". By opening the accompanying software packet(s), you acknowledge that you have read and accept the following terms and conditions. If you do not agree and do not want to be bound by such terms and conditions, promptly return the Book and the unopened software packet(s) to the place you obtained them for a full refund.

**1. License Grant.** JWS grants to you (either an individual or entity) a nonexclusive license to use one copy of the enclosed software program(s) (collectively, the "Software") solely for your own personal or business purposes on a single computer (whether a standard computer or a workstation component of a multi-user network). The Software is in use on a computer when it is loaded into temporary memory (RAM) or installed into permanent memory (hard disk, CD-ROM, or other storage device). JWS reserves all rights not expressly granted herein.

**2. Ownership.** JWS is the owner of all right, title, and interest, including copyright, in and to the compilation of the Software recorded on the physical packet included with this Book "Software Media". Copyright to the individual programs recorded on the Software Media is owned by the author or other authorized copyright owner of each program. Ownership of the Software and all proprietary rights relating thereto remain with JWS and its licensers.

**3. Restrictions on Use and Transfer.**

(a) You may only (i) make one copy of the Software for backup or archival purposes, or (ii) transfer the Software to a single hard disk, provided that you keep the original for backup or archival purposes. You may not (i) rent or lease the Software, (ii) copy or reproduce the Software through a LAN or other network system or through any computer subscriber system or bulletin-board system, or (iii) modify, adapt, or create derivative works based on the Software.

(b) You may not reverse engineer, decompile, or disassemble the Software. You may transfer the Software and user documentation on a permanent basis, provided that the transferee agrees to accept the terms and conditions of this Agreement and you retain no copies. If the Software is an update or has been updated, any transfer must include the most recent update and all prior versions.

**4. Restrictions on Use of Individual Programs.** You must follow the individual requirements and restrictions detailed for each individual program in the "About the CD" appendix of this Book or on the Software Media. These limitations are also contained in the individual license agreements recorded on the Software Media. These limitations may include a requirement that after using the program for a specified period of time, the user must pay a registration fee or discontinue use. By opening the Software packet(s), you agree to abide by the licenses and restrictions for these individual programs that are detailed in the "About the CD" appendix and/or on the Software Media. None of the material on this Software Media or listed in this Book may ever be redistributed, in original or modified form, for commercial purposes.

**5. Limited Warranty.**

(a) JWS warrants that the Software and Software Media are free from defects in materials and workmanship under normal use for a period of sixty (60) days from the date of purchase of this Book. If JWS receives notification within the warranty period of defects in materials or workmanship, JWS will replace the defective Software Media.

(b) JWS AND THE AUTHOR(S) OF THE BOOK DISCLAIM ALL OTHER WARRANTIES, EXPRESS OR IMPLIED, INCLUDING WITHOUT LIMITATION IMPLIED WARRANTIES OF MERCHANTABILITY AND FITNESS FOR A PARTICULAR PURPOSE, WITH RESPECT TO THE SOFTWARE, THE PROGRAMS, THE SOURCE CODE CONTAINED THEREIN, AND/OR THE TECHNIQUES DESCRIBED IN THIS BOOK. JWS DOES NOT WARRANT THAT THE FUNCTIONS CONTAINED IN THE SOFTWARE WILL MEET YOUR REQUIREMENTS OR THAT THE OPERATION OF THE SOFTWARE WILL BE ERROR FREE.

(c) This limited warranty gives you specific legal rights, and you may have other rights that vary from jurisdiction to jurisdiction.

**6. Remedies.**

(a) JWS's entire liability and your exclusive remedy for defects in materials and workmanship shall be limited to replacement of the Software Media, which may be returned to JWS with a copy of your receipt at the following address: Software Media Fulfillment Department, Attn.: *PMP: Project Management Professional Exam Review Guide, Second Edition*, John Wiley & Sons, Inc., 10475 Crosspoint Blvd., Indianapolis, IN 46256, or call 1-800-762-2974. Please allow four to six weeks for delivery. This Limited Warranty is void if failure of the Software Media has resulted from accident, abuse, or misapplication. Any replacement Software Media will be warranted for the remainder of the original warranty period or thirty (30) days, whichever is longer.

(b) In no event shall JWS or the author be liable for any damages whatsoever (including without limitation damages for loss of business profits, business interruption, loss of business information, or any other pecuniary loss) arising from the use of or inability to use the Book or the Software, even if JWS has been advised of the possibility of such damages.

(c) Because some jurisdictions do not allow the exclusion or limitation of liability for consequential or incidental damages, the above limitation or exclusion may not apply to you.

**7. U.S. Government Restricted Rights.** Use, duplication, or disclosure of the Software for or on behalf of the United States of America, its agencies and/or instrumentalities "U.S. Government" is subject to restrictions as stated in paragraph (c)(1)(ii) of the Rights in Technical Data and Computer Software clause of DFARS 252.227-7013, or subparagraphs (c) (1) and (2) of the Commercial Computer Software—Restricted Rights clause at FAR 52.227-19, and in similar clauses in the NASA FAR supplement, as applicable.

**8. General.** This Agreement constitutes the entire understanding of the parties and revokes and supersedes all prior agreements, oral or written, between them and may not be modified or amended except in a writing signed by both parties hereto that specifically refers to this Agreement. This Agreement shall take precedence over any other documents that may be in conflict herewith. If any one or more provisions contained in this Agreement are held by any court or tribunal to be invalid, illegal, or otherwise unenforceable, each and every other provision shall remain in full force and effect.

# The Best PMP Quick Reference Book/CD Package on the Market!

***Brush up on key PMP topics with hundreds of challenging review questions!***

- Two bonus PMP practice exams available only on the CD. Each question includes a detailed explanation.

Bonus Exam 1

You work for a specialized book publisher that publishes 35 new titles per year. Employees at your company have a specific process in place for managing each book project. They've found that managing most of the phases the same way across all the projects allows them to gain efficiencies they wouldn't have if they let each project manager use their own process. You're accepting your first assignment as the project manager for a new publication. You're responsible

A. Phase sequencing

B. Project management

C. Progressive elaboration

D. Program management

Your Answer:

Show Answer | Finish

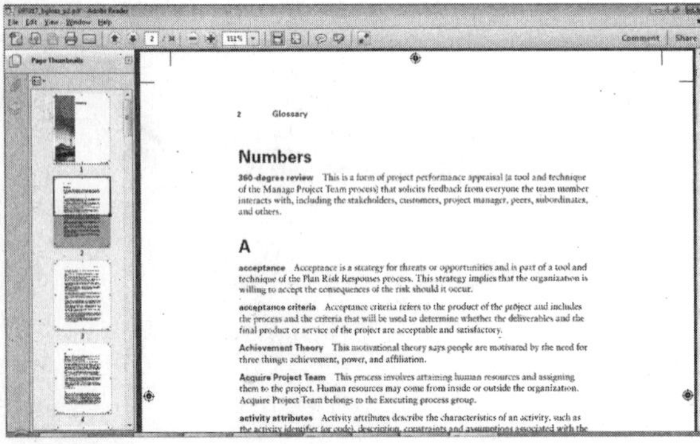

2 Glossary

## Numbers

**360-degree review** This is a form of project performance appraisal (a tool and technique of the Manage Project Team process) that solicits feedback from everyone the team member interacts with, including the stakeholders, customers, project manager, peers, subordinates, and others.

## A

**acceptance** Acceptance is a strategy for threats or opportunities and is part of a tool and technique of the Plan Risk Responses process. This strategy implies that the organization is willing to accept the consequences of the risk should it occur.

**acceptance criteria** Acceptance criteria refers to the product of the project and includes the process and the criteria that will be used to determine whether the deliverables and the final product or service of the project are acceptable and satisfactory.

**Achievement Theory** This motivational theory says people are motivated by the need for three things: achievement, power, and affiliation.

**Acquire Project Team** This process involves attaining human resources and assigning them to the project. Human resources may come from inside or outside the organization. Acquire Project Team belongs to the Executing process group.

**activity attributes** Activity attributes describe the characteristics of an activity, such as

***Use glossary for instant reference!***

- Search through the PDF of the glossary to find key terms you'll need to be familiar with for the exam.

***Use the electronic flashcards to jog your memory and prep last-minute for the exam!***

- Over 200 flashcards.
- Reinforce your understanding of key concepts with these flashcard-style questions.

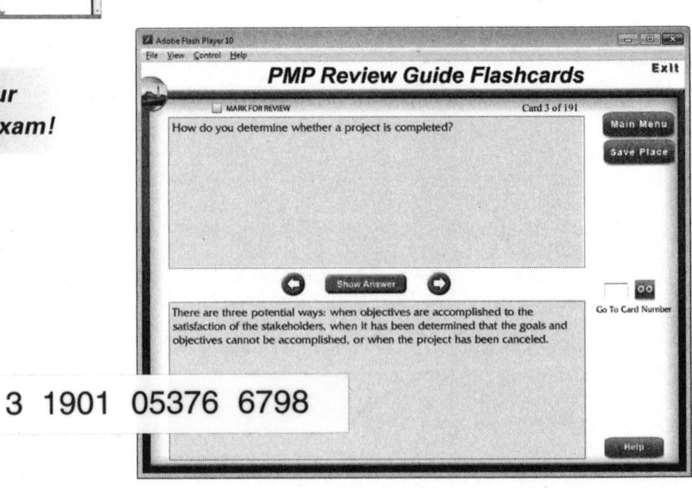


3 1901 05376 6798